KEEPING IT IN THE FAMILY

Keeping it in the Family

International Perspectives on Succession and Retirement on Family Farms

Edited by

MATT LOBLEY
University of Exeter, UK

JOHN R. BAKER
Iowa State University, USA

IAN WHITEHEAD
University of Plymouth, UK

Routledge
Taylor & Francis Group
LONDON AND NEW YORK

First published 2012 by Ashgate Publishing

Published 2016 by Routledge
2 Park Square, Milton Park, Abingdon, Oxon OX14 4RN
711 Third Avenue, New York, NY 10017, USA

Routledge is an imprint of the Taylor & Francis Group, an informa business

British Library Cataloguing in Publication Data
Keeping it in the family : international perspectives on
succession and retirement on family farms. -- (Perspectives
on rural policy and planning)
1. Family farms. 2. Family-owned business enterprises--
Succession. 3. Farmers--Retirement. 4. Estate planning.
I. Series II. Lobley, Matt. III. Baker, John.
IV. Whitehead, Ian
306.3'49-dc23

The Library of Congress has cataloged the printed edition as follows:
Lobley, Matt.
Keeping it in the family : international perspectives on succession and
retirement on family farms / by Matt Lobley, John Baker and Ian Whitehead.
p. cm. -- (Perspectives on rural policy and planning)
Includes bibliographical references and index.
ISBN 978-1-4094-0995-3 (hbk) -- ISBN 978-1-4094-0996-0 (ebook) 1. Family farms-
-Management. 2. Family-owned business
enterprises--Succession. I. Baker, John (John R.) II. Whitehead, Ian. III.
Title.
HD1476.A3L63 2011
338.1--dc23

2011049398

ISBN 13: 978-1-4094-0995-3 (hbk)

Contents

List of Figures

List of Tables

List of Contributors

Brian Ilbery is Professor of Rural Studies in the Countryside and Community Research Institute, University of Gloucestershire.

Damian Maye is a Senior Research Fellow in the Countryside and Community Research Institute, University of the West of England.

Dave Goeller is Deputy Director of the North Central Risk Management Education Center at the University of Nebraska, USA.

Elaine Barclay is Coordinator of the Bachelor of Criminology degree at the School of Behavioural, Cognitive and Social Sciences, University of New England, New South Wales Australia.

Ian Reeve is Senior Group Leader of the Institute for Rural Futures, University of New England, New South Wales, Australia.

Ian Whitehead is Associate Professor in Rural Property Management at the University of Plymouth, UK.

James Kirwan is a Senior Research Fellow in the Countryside and Community Research Institute, University of Gloucestershire.

John R. Baker is Attorney at Law and Administrator at the Beginning Farmer Center, Iowa State University, USA.

Joy Kirkpatrick is an Outreach Specialist at the University of Wisconsin Center for Dairy Profitability.

Julie Ingram is a Senior Research Fellow in the Countryside and Community Research Institute, University of Gloucestershire.

Linda Price is a Lecturer in the School of Planning, Architecture and Civil Engineering, Queen's University Belfast.

Matt Lobley is Co-Director of the Centre for Rural Policy Research at the University of Exeter, UK.

Mandi McLeod is an independent family business strategy advisor, specialising in business continuance and succession planning.

Nick Prince is a PhD student in the Countryside and Community Research Institute, University of Gloucestershire.

Peter C. Leach is a teacher, author and advisor to families.

Rachel Conn is a Masters student, Queen's University Belfast, employed by DARD (Department for Agriculture and Rural Development).

Roger Wilkinson is a rural social researcher with the Department of Primary Industries, Victoria, Australia.

Roslyn Foskey is a researcher at the University of New England, New South Wales, Australia.

Ruth Rossier is an Agricultural engineer at the Agroscope Reckenholz-Tänikon ART Research Station in Switzerland.

Tomohiro Uchiyama is Associate Professor of Graduate School of Bioresources at Mie University, Japan.

Forewords

Across the developed world, the future of agriculture depends on a supply of ambitious, talented and hard-working young people in order to sustain the supply of food and the broader range of environmental goods and services demanded by the public. If family farming is to survive in the longer term it is vital we retain a farming ladder which new entrants to agriculture are able to climb. If current trends continue, within a few years 'The Ladder' is in risk of being permanently broken with few opportunities to progress. Barriers to entry are often high, opportunities for genuinely new entrants are limited and the current generation of business principals are often reluctant, or uncertain, how to release their grip on the business in favour of a successor.

I, myself, benefited from the risks taken by others to help me get started. Giving an opportunity to a new entrant or someone with limited experience is a risk. If it had not been for a land owner and a bank manager who were both willing to take a risk I would never have made the first rung of the ladder. With appropriate advice, support and, if possible, mentoring the risk can be minimised and the investment really worthwhile.

Collectively, the contributors to this book have decades of experience of both research and practical advice on the transfer of the family farm business. The editors currently direct the *FARMTRANSFERS* research project and John Baker has drafted US legislation designed to facilitate the transfer of farm businesses.

It is vital that all farmers prepare properly for future succession, be it familial or non-familial. Successors need to be identified in a timely manner, offered appropriate training and given the opportunity to take control of the business (entirely or in part) in good time. This gives younger people the chance to make decisions for a longer-term future; it also gives them the experience of running a business that they will need when the incumbent business principal fully hands over to them.

This book makes an important contribution to the understanding of the challenges of farm succession and offers important clues about how to successfully plan the intergenerational transfer of a farm business.

Lord (Donald) Curry of Kirkharle KB CBE FRAgS

I have been in the business of financing farmers for over 30 years now. When I first started in 1980, the average age of American farmers was right around 50 years old. Today, that number is closer to 60. While farming and farming techniques have changed dramatically in the last 30 years, the one thing that does not seem to change is that the number of entry level farmers is declining and for those who do take the leap, it is getting progressively more difficult to navigate the technical and legal aspects of the profession and how to transition in to the business.

Not all entry level farmers have family support and, even for those who do, in many cases there is a lack of understanding of how to transition the business from one generation to the next. While I do not expect that any single event or educational opportunity can change the tide of the increasing age of our farmers overnight, I do believe that in each of these cases there will be a point at which a decision needs to be made on how to proceed.

Because a group of dedicated researchers and educators have made this issue their life's passion, information is now available to assist farmers to understand and execute a succession plan. Some of these farmers will choose to take an active role in deciding 'who farms their land', rather than leaving it up to the probate courts. Thankfully, this book will prepare both the withdrawing party, and the entry level party, with the information they need to make the transition as uneventful as possible.

Chris Beyerhelm, Deputy Administrator of Farm Loan Program for the Farm Service Agency, United States Department of Agriculture.

Preface and Acknowledgements

As the largest group of natural resource managers on the planet, farmers are at the interface of the changing relationship between humans and the environment. Typically organised around what might be considered the most basic of social units, for generations the family farm has survived wide-ranging exogenous challenges, frequently preserving the line of succession to the next of kin. Now as we face major questions about how we use land and the impact of our land use on the global environment, farming once again faces a challenging and uncertain future. This book draws on the experiences of farmers in Australia, New Zealand, North America, Japan and the EU to examine the special features of family farms and, in particular, the tradition of succession which has enabled them to continue to have such a strong presence in the world today.

Several of the chapters derive from research undertaken as part of the *FARMTRANSFERS* project initiated by the late Professor Andrew Errington, University of Plymouth. Those of us who worked with Andrew were inspired by his dedication to the study of family farming in general and farm succession and retirement in particular. That the *FARMTRANSFERS* questionnaire has been completed by many thousands of farmers in diverse parts of the world is testament to Andrew's infectious enthusiasm for the subject and we are grateful for all that he did.

In addition to the individual contributors to this book, many others have supported us and assisted in its preparation. In no particular order we are grateful to: Wendell Joyce (formerly of the Canadian Farm Business Management Council), Denis Chamberlain of the Royal Agricultural Society of England, Marc Falcone (formerly of the USDA), Michael Mack from Easton College, Alison Rickett from Fresh Start, Richard Soffe, Head of Duchy College Rural Business School, and Great Western Research and the Gruffydd Davies Legacy for valuable funding. Vicki Baker provided invaluable proof reading and editing services and Marilyn Stephen ably constructed the final manuscript. Finally, we would like to express our gratitude to our colleagues in the *FARMTRANSFERS* project and the many thousands of farmers who have completed a *FARMTRANSFERS* questionnaire.

Chapter 1
Succession and Retirement in Family Farm Businesses

Matt Lobley and John R. Baker

Introduction

The long predicted demise of the family farm (e.g. Marx 1976) has proved to be somewhat exaggerated. Rather than being subsumed by capital, family farming is characterised by tenacity and persistence. This does not mean that family farms are unchanging, far from it; and the explanations of the persistence of family farming are both varied and contested (see Calus and Huylenbroeck 2010, Brookfield and Parsons 2007). Now, following three decades in which the dominant policy discourse in the developed world has been one of land surplus, set aside and alternative land use (Lobley and Winter 2009), family farmers are faced with responding to the challenge of the 'perfect storm' of rising demand for food at the same time as commentators envisage an increasing role for energy production from farmland, increased demand for water and adapting to and mitigating climate change (Godfray et al. 2010). Globally, price volatility is an on-going feature of agricultural markets (OECD 2010), whilst agricultural land prices continue what appears to be an inexorable rise. For instance, British agricultural land prices reached an all-time high in 2010, stimulated by strong demand from commercial farmers and falling supply (RICS 2011). Similarly, in the United States the value of farmland has been on an upward trend for decades (USDA 2010).

A major reason behind strong and rising agricultural land prices has been the entitlement to agricultural support payments which have effectively become capitalised into the value of farmland (Trail 1979). As land values increase it is difficult for new farmers to finance the purchase of land and for farms to expand to bring a successor into the business. In situations where the farm expands through the purchase of land using borrowed funds, it can increase the risk of the failure of the farm business due to either a crop or market failure. Further, such acquisition may also limit the ability of the business to acquire new equipment or add a new enterprise to provide the needed income for the successor.

Family farms face challenging times. Many, however, will still hope to see the business passed on to a familial successor. Indeed, a major concern for many family businesses is to pass control of a sound and often improved business to the next generation (Milton 2008, Sharma et al. 2003), a process which often triggers new phases of business development and which, when completed successfully, can confer competitive advantage on the family business (Sharma et al. 2003). This desire to secure family leadership of the business across the generations is particularly important given that family businesses tend to be the dominant business model in most developed economies (Zachary 2011, Milton 2008, Aldrich and Cliff 2003). In the case of the family business, their unique characteristics are such that in many cases the successor is a child of the incumbent manager, raising issues of shifting intra-family power dynamics and individual identity. It also means that as well as physical business assets, intangible assets including firm-specific tacit knowledge are transferred to the new 'chief executive' (Lobley et al. 2010, Danes et al. 2009, Milton 2008). The early exposure of children to the firm's working practices can produce deep levels of firm-specific tacit knowledge which can give 'family employees of family firms the potential to have deeper levels of firm-specific knowledge than employees of nonfamily firms' (Danes et al. 2009: 202). In an agricultural context, in addition to succeeding to the farm, the successor also benefits from the transfer of skills and, frequently, a detailed knowledge of the home farm, its micro-climate and idiosyncrasies.

When the fact that most farming businesses are operated by families (Brookfield and Parsons 2007) is taken into consideration, it can be argued that intergenerational farm transfers are a fundamental aspect of the sustainability of family farming systems. Intergenerational transfer is the process that reproduces the family farm and farms that are not capable of being transferred to a succeeding generation are arguably merely long-term temporary operations rather than sustainable family farms. In this context the transfer of business ownership and control to the next generation is arguably one of the most critical stages in the development of the business. This may involve the transfer of the 'home farm' to a successor (or multiple successors) or it may involve the transfer of the necessary capital to establish a new farm business. In this way it is possible to distinguish between succession to the farm and succession to the occupation of farming (Lobley et al. 2010). *Intergenerational Succession* then is the process, stretching over a period of time, of transferring managerial control and other intangible assets such as firm (farm) specific knowledge. The mirror image of succession is *retirement*, again often a gradual withdrawal from physical labour and managerial control by the older generation. The analogy of 'handing over the reins' has been used to describe the stages of the succession process in farming (Errington and Lobley 2002), while Dyck et al. (2002) used the image of a 4 × 100 relay race in their 'passing the baton' analogy. Important stages in ensuring completion of the race are 'sequence', i.e. ensuring that the successor obtains the appropriate skills and experience in a step-wise manner, and 'timing', i.e. ensuring that the leadership baton is effectively passed to the successor as the incumbent withdraws from his or her leadership role (Dyck et al. 2002). Finally, *inheritance* denotes the legal transfer of ownership of business assets (including land). Whilst conceptually separate, these processes are linked and the timing and smoothness of the process can have considerable implications for the farm business as well as the individuals involved in that business.

The process of intergenerational succession can be a time of considerable stress in the family business (Milton 2008, Burton and Walford 2005) and there is much evidence of the impacts on the successor and the business when the older generation cannot bring themselves to fully let go of the reins (Errington and Lobley 2002, Potter and Lobley 1992). Often the incumbent will have spent much of their life, not just their career, learning about the business, identifying with the business, developing the business, making sacrifices for the business and taking risks for the business, and the prospect of withdrawing from the leadership role can be both challenging and frightening as it may be associated with fears of 'the loss of power, status, or personal identity' (Sharma et al. 2003: 671. Also see Milton 2008). Both the successful transfer of the business and the 'failure' of succession can have a powerful influence on the development trajectory of a family business, whether it is a farm business or not (Milton 2008, Calus, Huylenbroeck and Lierde 2008, Mishra and El-Osta 2008, Potter and Lobley 1992, Boehlje and Eidman 1984, Harl 1972). The identification of a successor can act as a trigger for business development and existence of a successor can provide a powerful motivation for on-going investment in the business even into the old age of the retiring farmer (Potter and Lobley 1992). Failure to achieve smooth and timely succession can lead to the depletion of capital assets and ultimately, the failure of the business (Milton 2008, Potter and Lobley 1992).

Despite the importance ascribed to succession in family business studies it has been argued that very little is known about the actual process of succession (Dyck et al. 2002). In the context of farming, a combination of an ageing farm population and significant barriers to the entry of new and young farmers are common characteristics of the agricultural sectors in many developed countries (ADAS et al. 2004). Consequently, a clearer understanding of the process of succession and retirement on family farms is crucial in order to appreciate the ability of family farms to renew themselves through intergenerational transfer, and to identify the type of interventions that can help facilitate timely intergenerational intrafamilial succession or the movement of new entrants into agriculture. Against this background this chapter introduces the subject matter of the book and several of the concepts and issues discussed in subsequent chapters. It considers the importance of the family farm and how ideas about the family farm have influenced thinking about the relationship between farmers, farming, state, society and culture, particularly in Europe and the United States. The chapter then goes on to explore why succession remains important in contemporary family farming systems, and discusses some of the direct and indirect impacts of succession on the farm business. It also considers the implications of the 'failure' of succession and the potential role for policy makers to intervene to facilitate and/or speed up the twin processes of succession and retirement. Given that several chapters in this book draw directly on the *FARMTRANSFERS* methodology, this chapter also briefly outlines the origins and history of the *FARMTRANSFERS* approach.

Family Farming in Policy and Society

The family farm has long held an important place both in policy and society. For instance, in Western Europe during the formative years of the CAP (Common Agricultural Policy), family farming was typically seen as the bedrock of European rural society and rural economy. This was reflected by the founding fathers of the European Community who, in the 1965 Spaak report (which formed the basis on which the EEC was established), recognised the importance of the 'social structure of agriculture based on the family farm' (Fennell 1987: 5). Then, in the year following the signing of the Treaty of Rome, the 1958 Stresa Conference established the principles of the CAP, issuing a general resolution that agriculture was both an integral part of the European economy and an essential factor in social life. Moreover, it was argued that 'given the importance of the familial structure of European agriculture and the unanimous wish to safeguard this character, every effort should be made to raise the economic and competitive capacity of such enterprises' (quoted in Fennel 1987: 11). Thus, in the early years of the EEC, family agriculture was positioned as the 'economic engine' driving rural development and its economic position was coupled with an important social role. Even at the beginning of the 21st century, then EU Agriculture Commissioner Franz Fischler argued that the core of the EU's farming system is the family farm and that it deserved continued support (Fischler 2001).

Ideas about the family farm have also played an important role in the persistence of the North American agrarian ideal. Founding father of the USA, Thomas Jefferson argued that farmers were the most valuable citizens and 'the most independent, the most virtuous, and they are tied to their country and wedded to its liberty and interests by the most lasting bonds' (Jefferson 1785, quoted in Browne et al. 1992). Jefferson's argument was that an independent yeomanry was the basis of democratic society and he emphasised the relationship between farming, citizenship and stability. Writers in the 19th century added a new dimension by stressing the moral and spiritual benefits of farm work. Thus, the 'myth' of the family farm combined 'Jefferson's hardworking yeoman with a legendary superiority stemming from the prevailing Protestant work ethic of handwork as a measure of moral worth' (Bonnen and Browne 1989: 12).

Contemporary agrarian writers such as Wendell Berry draw on these notions to uphold small scale family farming, arguing that '… the small farm of a good farmer, like the small shop of a good craftsman, give work a quality and a dignity that it is dangerous for … human work to go without' (Berry 2009: 34). To Berry, the family farm is 'part of one's own humanity' (page 31), supporting a superior quality of life and higher moral and spiritual values than industrial society. This view is summed up by Paarlberg (1980: 185) who states that for supporters of family farms, the loss of such farms would imply the loss of a valuable way of life: 'the family farm represents in the minds of many, an idealised form of preindustrial living, the son apprenticed to the father, living close to nature and producing the most needed product of all … .' The romanticised view of family farming, extolling craftsmanship and creativity can also be found among British writers: H.J. Massingham for instance, writing about a small farm in Kent says 'They farm in precisely the same way as a poet writes a sonnet or a sculptor carves from the block. They are in the most definite application of the term artists who assemble the materials of their craft into a creative unity' (Massingham 1948, *An Englishman's Year*, quoted in Keith 1975). Interestingly, Emmerson also likened a good farmer to the true poet. In the case of Britain, it has been argued that a group of farming writers (or writing farmers) such as George Henderson (author of *The Farming Ladder*, originally published in 1944) were in part responsible for promoting the virtues of small scale family farming (Winter and Lobley 2006).

In many ways the American view of the family farm is not unlike the European perspective. Indeed, it has been argued that the notion or the ideal of the family farm has near universal appeal (Frances 1994). That said, although the family farm is generally held in high regard (Gasson 1987) there is no general agreement on what is meant by family farming, although in common usage family farm tends to imply a smaller farm. In this context, Gasson and colleagues deserve to be quoted at length when they observe that

> … policymakers and their advisors may be subconsciously influenced by certain stereotypes … There would be less confusion if politicians and policy makers would specify 'the small farm' or the 'family worked farm' if this is what they mean, rather than introduce references to 'family farming' which, while having a certain emotional appeal, may not mean very much. It is after all, a great mistake to be influenced by an over-romantic view of the family farm (Gasson et al. 1988: 35).

Within the UK, the individual devolved governments adopt distinctly different attitudes to the importance of the family farm. For instance, in England, the approach taken by the Department of the Environment, Food and Rural Affairs (Defra) is decidedly agnostic and does not extol the virtues of any particular size structure of farms and does not privilege 'family' farms above other forms of farming. While Defra remains silent on the topic of preferred types of farming structure, in neighbouring Wales the National Assembly appears to firmly favour the family farm arguing that 'the family farm defines the character of Welsh rural society, and it's sense of identity. The numbers directly and indirectly involved in farming make a crucial contribution towards sustaining rural communities' (National Assembly for Wales 2001). Beyond Europe, strong connections are also made between sustaining family farms and rural sustainability. As we have seen, in the United Sates, family farmers have long held a special place in the nation. Despite the transformation of American agriculture, it has been argued that family farms are

> … still a vibrant and necessary part of society … the contribution of farm businesses to the rural economy, the environment and, indeed, to society as a

> whole should not be ignored. Those who live next to the land and toil upon it are more likely to be interested in their community, their environment and their society … (Baker 2005).

This, of course, still begs the question of what is a family farm?

Whilst it is not the intention of this chapter to engage in an extended discussion of the definition of the family farm, it is important to explore some of the characteristics of such farms and their approaches to their definition. Broadly speaking, there are two ways of approaching the issue of definition. The first is to employ an *ideal type* definition, an approach widely adopted in academic research. The purpose of such definitions is to raise awareness of significant relationships, e.g. between the farm family and the farm business and to highlight their consequences (Gasson and Errington 1993). The ideal type approach aims to highlight 'issues which can be understood and discussed at a general level, rather than to construct a precise, watertight definition' (Gasson 1993: 11). This approach focuses largely on the different processes occurring within different groups of farms and farmers, often defined not in terms of easily quantifiable agricultural and economic factors but in terms of attitude, outlook and propensity to behave in a certain way. The use of ideal types and the development of typologies though should not be regarded as an end in itself but as a methodological tool which provides a link between theory and practice (Whatmore et al. 1987). Although this approach can be used to inform policy, the type of definition required to enact policy must be amenable to measurement and quantification. For example, devising cut off points based on accepted, understood and measurable criteria is often a necessary exercise. The purpose of such an *operational definition* is to select or exclude farms/farmers, often through the use of (arbitrary) cut off points (see Hill 1993) as opposed to the ideal type approach which is more concerned with grouping farms and farmers together with a view to analysing significant relationships and processes.

The United States Department of Agriculture's (USDA) Economic Research Service (ERS) currently defines a family farm as 'any farm where the majority of the business is owned by the operator and individuals related to the operator by blood or marriage, including relatives who do not reside in the operator's household' (Hoppe and Banker 2010: 2). Based on this definition, 98 per cent of farms in the USA are considered to be family farms. Perhaps more useful is the USDA/ERS farm typology (see Table 1.1) which not only distinguishes between small and large family farms but identifies a range of other farm types and subdivisions, including retirement farms and residential/lifestyle farms of the type that are characteristic of many parts of the British countryside but for which data on their distribution is not easily available. Using this typology, in 2007 63.5 per cent of US farms were retirement or residential farms and these small family farms accounted for 5.8 per cent of the value of US agricultural production. On the other hand, very large family farms and non-family farms, which together account for just 7.4 per cent of all farms, were responsible for 71.4 per cent of the value of agricultural production (Hoppe and Banker 2010).

Table 1.1 USDA economic research service farm typology

Small family farms (gross sales less than $250,000)	**Large-scale family farms (gross sales of $250,000 or more)**
Rural-residence family farms: **Retirement farms**. Small farms whose operators report they are retired. **Residential/lifestyle farms**. Small farms whose operators report a major occupation other than farming.	*Commercial family farms*: **Large family farms**. Gross sales between $250,000 and $499,999. **Very large family farms**. Gross sales of $500,000 or more
Intermediate family farms: **Farming-occupation farms**. Small family farms whose operators report farming as their major occupation. **Low-sales farms**. Gross sales less than $100,000. **High-sales farms**. Gross sales between $100,000 and $249,999.	**Non-family farms** Any farm not classified as a family farm, i.e. any farm for which the majority of the farm business is not owned by individuals related by blood, marriage, or adoption.

Source: Hoppe and Banker 2010. Reproduced with permission of USDA/Economic Research Service.

In Europe, Hill (1993) developed an operational definition in his study of family, or more precisely, *family-worked* farms. Controversially for some, he argued that the family-worked farm is not necessarily synonymous with the small farm as some small farms are run mainly with hired labour. This approach, however, runs counter to what is perhaps the more accepted view that a family farm is run by and for the family, the family having to provide *some* labour (managerial or manual), as well as some business capital. In the broader sphere of family business studies, family firms are usually 'defined by criteria or combinations of criteria including family ownership, management by a family member, operational involvement of family members, and family member involvement across generations' (Rogoff and Heck 2003: 560). Milton (2008) suggests that family members should have active involvement at a strategic level, while Zachary (2011) reviews a range of approaches to defining family businesses involving (among other factors) varying degrees of family labour commitment. Within agriculture, researchers have stressed that the presence or absence of hired labour, as opposed to family labour, is often a function of the family life-cycle and enterprise mix and is not on its own sufficient basis for distinguishing between family and non-family farms (Gasson and Errington 1993). Although definitions of the family farm vary, there is an emphasis on the family as 'the risk taking manager' (Galeski and Wilkening 1987: 303), suggesting that even on small farms where certain tasks may be carried out by contract labour, these farms can be regarded as family farms as the family is still the risk talking manager. Gasson argued that 'use of family labour has become a less distinctive feature of farm organisation and therefore a less relevant criteria for defining the family farm', although there has been vigorous debate over this issue (e.g. Djurfeldt 1996, Errington 1996).

Long-time observers of family farming, Ruth Gasson and Andrew Errington, developed an ideal type definition based on the following multiple criteria:

1. Business ownership is combined with managerial control in the hands of business principals.
2. These principals are related by kinship or marriage.
3. Family members provide capital to the business.
4. Family members, including business principals, do farm work.
5. Business ownership and managerial control are transferred between the generations with the passage of time.
6. The family lives on the farm.

(Gasson and Errington 1993: 18)

While this can be seen as a comprehensive definition in terms of the factors included, this approach does not suggest that family farms are a certain size in terms of sales, labour or area. The criterion of intergenerational transfer is also of particular interest. Definitions of family businesses often include an intergenerational dimension (Zachary 2011). However, there is widespread evidence that small family farms are less likely to attract a successor (Lobley et al. 2010, Uchiyama et al. 2008, Glauben et al. 2004). Thus, small farms without a successor would not meet Gasson and Errington's criteria of the family farm. This also begs the question of whether a farm that has previously been transferred between generations but fails to attract a successor at some point would cease to be a family farm, or when it is purchased by a new family does it not become a family farm until transferred between the generations again? For Berry (2009) the longevity of the connection between family and farm is one of the defining features of the family farm and contributes to a shared family knowledge of errors to avoid and solutions to problems. It is suggested that a farm that is newly acquired is *potentially* a family farm, depending on the intentions of the family occupying it. A farm that has already been passed down through generations of a family '... may rank higher as a family farm than a farm that has been in a family for only one generation; it may have a higher degree of familiness or familiarity than the one generation farm' (page 32). On the other hand, in their definition of a family farm, Brookfield and Parsons (2007: 14) argue that '... it is much less diagnostically significant that the farm be handed down through generations', instead arguing that the defining characteristics should be that the farm is managed by the 'family group', that they are smaller than average for their country and time, and that half or more of the labour requirement of the farm is provided by the family. To be fair to Gasson and Errington, they argued that the achievement of intergenerational succession was one of the less important defining features of a family farm, but, in line with Berry, recognised that the *desire* to pass the farm on was an important characteristic of the farm family business.

In part, this discussion of the meaning of the family farm is designed to illustrate that the search for a definition of the family farm that would garner widespread agreement and support is futile. Researchers and commentators will continue to employ different approaches, drawing on different perspectives, different academic disciplines and different data. For current purposes it is the close link between family and business that is most important. It is the 'family-ness' of the organisation of the business that is important, rather than minimum or maximum criteria for farm size, turnover, or family labour contribution. In recognising the importance of the family, Gray's (2000: 220) notion of 'consubstantiality' of family, farm and place provides a powerful intergenerational driver: 'keeping the farm in the family is taken literally. It includes not just the explicit legal transmission of the farm's assets and operational control to a successor but also the implicit reproduction of a family's self-defining and consubstantial sense of place on the farm'. This has close links to Björnberg and Nicholson's (2008: 3) notion of emotional ownership; 'the idea that the business is, in some sense, part of who you are as a person'. Although not directly a cause of business success, emotional ownership and the deployment of 'family capital' – the commitment and participation of family members (Björnberg and Nicholson 2008) can confer advantages on the family business. Indeed, within family business studies the notion of 'familiness' has emerged relatively recently and is typically described as the unique bundle of resources resulting from the interaction of family and business, the sum of which may be greater than the individual parts (Danes et al. 2009, Pearson et al. 2008, Milton 2008). Familiness is a broad concept that is generally thought to offer competitive advantage to family firms (Pearson et al. 2008), where vision and commitment to the business are deeply embedded in family history. Such strong personal commitment and consubstantiality can have profound effects on the individuals involved and not always in a positive manner. As Price (2010: 88) observed through multiple interviews with farming men at various stages of their lives 'it was possible to reveal the deep attachment to the family story they feel bound to continue. It is not difficult to see how such an intense sense of duty to the past, present and future can restrict life-choices and influence relationships with other family members around this imperative'. Milton (2008) also points to a darker side of familiness in terms of the destructive tendencies of some relationships, while Pearson et al. (2008) recognise that 'too much' familiness can, for instance, result in a closed minded approach to new ideas.

The Significance of Succession in the Family Farm Business

Although notions of the family farm have long influenced thinking about the role of agriculture in society, the precise meaning of the term remains contested. As a form of business organisation the family farm has proved both adaptable and resilient and the familiness of the business can confer a powerful joining of family, business and place that can exert a profound influence on the individuals involved. Given the widespread importance of family farming, it can be seen that intergenerational farm transfers are a fundamental aspect of the sustainability of family farming systems.[1] Moreover, many commentators argue that sustaining family farms contributes to the broader sustainability of rural communities (e.g. Ramos 2005). A number of authors (e.g. Appleby 2004, Burton et al. 2005, Parry et al. 2005) have argued that a decline in the number of 'traditional' family farms is associated with a decline in the provision of public environmental goods and in the organisations and institutions that are the mainstay of rural life. Clearly, decisions made (or sometimes avoided) about the management of family farms have implications for rural economies, rural communities and the environment. Farm family businesses face a range of complex drivers for change and while much attention has focused on macro analysis of global processes of trade liberalisation and seemingly endless rounds of agricultural policy reform, the farm household is also important because it is here that the effects of other drivers of change are mediated, as well as being a source of internal farm household drivers of chnage. Indeed, a considerable body of evidence (e.g. Potter and Lobley 1996, Gasson and Errington 1993, Bryden et al. 1992) suggests that family events and processes such as births, marriage, ageing, succession and retirement can influence reaction to changes in the external environment and can trigger restructuring in agricultural businesses. Similarly, Aldrich and Cliff's (2003) 'family embeddedness perspective' emphasises how family transitions (including childbirth, marriage, divorce, retirement and death) impact on both family and business in a variety of ways.

In simple terms, intergenerational succession is important because it represents an integral facet of the family farm. The intergenerational and intra-familial transfer of farms can be a source of great strength. Intergenerational succession represents the renewal of the family farm and can potentially act as a helpful corrective in addressing the apparent increasingly aged population of principal farmers.[2] In the UK (and many other countries) families are responsible for most farms and much farmed land. For example, a recent unpublished University of Exeter survey of 1543 farmers in South West England found that 81.5 per cent operated 'established family farms' (i.e. those who are at least the 2nd generation of their family to be farming the same farm or nearby farm), and were responsible for managing 81.5 per cent of the area covered by the survey. Sometimes family occupancy of the farm or local farmland was extremely lengthy and 28 per cent of established family farmers could trace their family's occupancy of the farm to 1900 or earlier (known as Century Farms in the USA). Not surprisingly, many (63 per cent) had been responsible for their farm for at least 20 years, although a significant minority (15 per cent) had assumed responsibility in the last ten years.

---

1 This is not to suggest that intergenerational succession is or should be the only means of entry in to farming. Far from it, the 'new blood' effect of entrants from outside the agricultural sector has long been recognised (e.g. Northfield 1979).

2 To an extent the frequently quoted figures demonstrating the high average age of British farmers is misleading. 'Official' figures are based on the age of the registered holder of the holding. In many cases this will be an older person but the individual responsible for the day to day running of the business will often be much younger.

Fewer than 10 per cent were new entrants in the strictest sense that they were the first generation of their family to farm in the locality and had not previously farmed elsewhere. This confirms previous research which indicates that the main entry route into farming in England remains intergenerational transfer within a family (ADAS et al. 2004).

In Australia, despite falling rates of succession, some 94 per cent of farms remain family owned and operated. Many farmers can trace their family's occupation of the farm back to three generations or more and there is evidence of a strong 'rural ideology' that prioritizes passing the farm on (Barclay et al. 2005 and Chapter 2 of this book). Patterns of ownership in the United States are similar to those found in the UK and Australia. As we have seen, most farms in the United States are considered to be family farms and those farm families own 93 per cent of all farm assets (Hoppe and Banker 2010). Evidence suggests that longevity of land ownership varies regionally in the US (Lobley et al. 2010) although in Iowa, the average length of ownership of family farms is some 83 years (Iowa Rural Life Poll 2009).

The process of intergenerational transfer is not just about the renewal of the family farm and keeping the name of the land. In the process of handing over managerial control, sometimes over a period of many years, the associated transfer of farm-specific, or 'soil-specific' human capital can confer an advantage on an intergenerational successor (Laband and Lentz 1983). In a developing county context, Rosenzweig and Wolpin (1985) argued that the accumulation of farm-specific knowledge had the effect of increasing labour productivity and made the farm and its land of more value to the farming family than a potential outside purchaser. Agriculture in developed countries is also a knowledge-rich undertaking and farm-specific knowledge is arguably becoming more important with a market turn which favours quality. It has been argued that farm and location specific knowledge is important in the production of quality foods, that intrafamily succession can help by transmitting specific knowledge from one generation to another (Corsi 2004) and that farm-specific human capital increases the value of transferred physical assets (Glauben et al. 2004). Kennedy (1999: 133) reasoned that 'family members may have accumulated farm-specific capital, in the sense of productive skills and knowledge peculiar to the home farm, which gave them an efficiency advantage over non-kin'. In addition, the highly detailed and locally specific knowledge associated with successful intergenerational transfers can prove vital for effective environmental management and, thorough engendering a sense of intergenerational accountability, can position farmers as interpreters and exemplars of local history, nature and culture (see Burton et al. 2005).

In family farming businesses, where the number of individuals involved in any one business is relatively small, the importance of human capital, which includes the skills, talents and tacit knowledge that has been developed by the farmer in working a particular area of land, should not be underestimated, particularly in the context of undertaking locally appropriate sustainable land management practices. Generally speaking, farmers draw on two forms of knowledge: farming practice in general (i.e. codified or standardised knowledge) and tacit, farm-specific knowledge, such as a detailed knowledge of the home farm, its individual fields and micro-climate and its idiosyncrasies. In the case of agriculture, Laband and Lentz (1983) have argued that farm-specific knowledge is of far greater value than firm-specific knowledge in other sectors.[3] Moreover, the larger the farm business, the more vital managerial skills and knowledge become, as opposed to the physical assets themselves, and the more important the smooth and timely transfer of such skills. Thus, Winter (1995) has highlighted the importance of farmers' knowledge transfer in relation to running the farm business under the circumstances of farmer isolation, whilst Dumas et al. (1995) have defined succession as the process of transferring managerial know-how.

The impact of succession, however, extends beyond the transfer of knowledge. Evidence from the USA and Europe suggests that farm business performance and farm development can be influenced by succession (e.g. Calus et al. 2008, Mishra and El-Osta 2008, Potter and Lobley 1992, Boehlje and Eidman 1984, Harl 1972). Such influences can operate in a number of ways. Potter and Lobley (1996) identified three principal effects associated with succession and retirement. The first is the 'succession effect' which refers to the impact of the *expectation* of succession on the farm business. Farms may be developed over a long period in order to provide a business capable of supporting two generations or to yield sufficient capital to establish successors on separate holdings. The succession effect can operate from close to the time of the birth of the first 'potential successor' although it is more likely to be felt when a successor indicates their intention to follow the occupation of farming:

> … once married and with children of my own, my ambitions became stronger to provide a good standard of living and improve the value of the farm. Once both sons definitely wanted to come home then expansion and improvement plans came to fruition.
>
> Because I'm in partnership with two sons we're in full swing, we are going forward. If I was on my own things would be very different. I wouldn't have bought the new farm for a start (Quoted in Potter and Lobley 1996).

Using data from the 2001 US Agricultural Resource Management Survey, Mishra and El-Osta (2008) also identified a positive association between farm capital stock and succession decisions on US farms and Calus et al. (2008) found that the value of total farm assets was significantly higher on Belgian farms where a successor was present.

3 The authors show that rates of intergenerational succession are much higher in farming than in other self-employed occupations and conclude that this must be an indication of the greater value of tacit knowledge in farming than in other self-employed occupations.

Farms with a successor present are much more likely to have a history of significant capital investment and expansion than farms lacking a successor. Such indications of an *association* between the likelihood of having a successor and the size and value of the farm raise interesting questions regarding *causality*. For instance, do farms grow because they have a successor or do larger farms attract a successor more easily? The concept of the *succession effect* would suggest the former, with growth and investment then reinforced by the *successor effect*. The concept of the 'successor effect' (Potter and Lobley 1996) refers to the impact of the successor themselves as they gradually (or sometimes rapidly) assume managerial control. Successors often return from a period of agricultural training with new ideas and an innovative approach to the business. The extent of their impact will be influenced by how rapidly they ascend the 'succession ladder' (see Errington and Lobley 2002), although Potter and Lobley (1996) report that over 9 per cent of all capital investment, 6 per cent of all land purchase and 8 per cent of all major enterprise change in a sample of 504 farms across Britain over a 30-year period took place within a year of the successor's return to full-time work on the farm.

Finally, towards the end of a farmers' career the 'retirement effect' (Potter and Lobley 1996) can be identified and is most pronounced where succession has been ruled out. In these cases farm operators frequently disengage or even withdraw from agriculture. This may involve down-sizing to reduce work load, letting or selling land and/or farming remaining land less intensively. In some instances, these farmers are effectively consuming their capital assets as they progressively liquidate farm assets to provide an income as part of a gradual process of leaving farming. For example, evidence from Belgium indicates that older farmers without successors begin to disinvest and that total asset values can decline towards liquidation levels (Calus et al. 2008).

The twin processes of succession and retirement can have a profound impact on both the family business as an entity and the individual members of the family directly involved in the business, or those with a stake in it. Paradoxically, some of the strengths of the family business; the close alignment of self and business, a willingness to self-exploit for the sake of the family and business can make the transition of succession a difficult time which may itself be exacerbated by an unwillingness to plan for a transition which, by definition, involves shift in intra-family power relations and also possibly a perceived loss of status on the part of the older generation. Consequently, it is not surprising, perhaps, that initiatives exist in several countries that are designed to facilitate the transition of the business between the generations. A number of such initiatives are discussed in subsequent chapters.

The Structure of the Book and a Note on Methodology

Despite the importance of intergenerational succession, both as a mechanism for transferring tangible and intangible farming assets and as an objective in itself, relatively little attention has been given to the pattern and process of farm succession. Moreover, much less attention has been paid to international comparisons of the succession process. Those studies that have been undertaken tend to have been conducted within a single country. Against this background, this book considers why intergenerational succession remains important in contemporary agriculture. Uniquely, it brings together contributions from academic researchers as well as practitioners involved in designing and implementing initiatives to facilitate the intergenerational transfer of family farms. Chapters 2–6 identify differences in rates and patterns of succession in a diverse set of farming, political, economic and cultural contexts and address some of the impacts of the twin processes of succession and retirement on the farm business in the process of considering some of the socio-economic factors associated with the process of succession. Chapters 7–11 place greater emphasis on the analysis of policy approaches for facilitating entry and exit from agriculture, with Chapter 7 considering non-family succession routes of entry into agriculture in England and Wales. Although the focus of this book is firmly on intergenerational succession in family farming, Chapter 12 offers insights from family business studies and illustrates how farm family business and their advisors can learn from other sectors. Finally, Chapter 13 draws the threads of the discussion together and considers some of the strengths and challenges facing farming families as they look to transfer the businesses to a new generation of business principals who will inevitably confront a new set of business challenges.

Several of the following chapters draw on *FARMTRANSFERS*, an international collaborative project based around a common research design which gathers information on succession and retirement plans and the speed with which a range of decisions are shared with and ultimately delegated to the successor. The objectives of *FARMTRANSFERS* are to:

1. Confirm the elements of farm succession plans.
2. Establish whether or not there is an identifiable career ladder in farm business successions.
3. Compare the patterns of succession in the participating Countries, States, Provinces and/or Territories.
4. Determine educational needs of farm business owners regarding succession.
5. Create a data archive that is available to research collaborators.

FARMTRANSFERS employs a copyrighted questionnaire developed by Professor Andrew Errington, which draws on previous work by Errington (1984) and Hastings (1984) on succession and the delegation of farm decision-making. Data is collected through a postal questionnaire designed to capture a range of information on plans for succession and retirement, information sources used, expected sources of retirement income and detailed information on the delegation of decision-making responsibility between the principal farmer and his/her successor(s), as well as basic background information on the farm. The survey has now been replicated in ten countries and in seven states in the United States (see Table 1.2) and the questionnaire has been completed by over 15,600 farmers. *FARMTRANSFERS* is currently directed by the three co-editors of this book.

Table 1.2 Replications of the *FARMTRANSFERS* survey

Australia (2004)	Japan (2001)
Austria (2003)	North Carolina (2005)
California (Humboldt County, 2004)	Pennsylvania & New Jersey (2005)
Canada (Ontario & Quebec, 1997)	Poland (2003)
England (1991, 1997)	Switzerland (2003)
France (1993)	Tennessee (2010)
Germany (2003)	Virginia (2001)
Iowa (2000, 2006)	Wisconsin (2006)

By adapting a common questionnaire to investigate patterns of farm succession, retirement and inheritance in a diverse range of social, cultural and economic contexts, *FARMTRANSFERS* has developed a unique database of comparable information. However, in methodological terms, the approach adopted by *FARMTRANSFERS* has both strengths and weaknesses. In addition to the limitations imposed by the standardised postal questionnaire format, the wide range of social, cultural and economic differences in the different countries and US states participating in *FARMTRANSFERS* means it is necessary to adapt each replication slightly. Such modifications to the questionnaire are made with the agreement of the project directors. Individual replications of the survey can also vary considerably in terms of the year of the survey (*FARMTRANSFERS* now spans two decades) and sample size. That said, the *FARMTRANSFERS* approach yields a range of (largely quantitative) data relating to the pattern, process and speed of succession and retirement which provides a firm base for future inquiries utilising different methodologies. Moreover, it allows for an international comparison of the results which is not possible using other data sets. Consequently the data is invaluable in order to identify widespread elements of succession plans, to establish the advisory needs of farm business owners as they contemplate succession retirement and inheritance, to compare succession patterns internationally, and to create a resource useful to farm business operators for future succession activities.

Our approach to *FARMTRANSFERS* is that it may not be methodically perfect but, along with the other contributions to this book, it does help illuminate a complex, fascinating and important aspect of family business life.

References

ADAS, University of Plymouth, Queen's University Belfast and Scottish Agricultural College, 2004. *Entry to and Exit from farming in the United Kingdom*. ADAS Consulting Ltd, report to Department for Environment, Food and Rural Affairs, London.

Aldrich, H. and Cliff, J. 2003. The pervasive effects of family on entrepreneurship: toward a family embeddedness perspective. *Journal of Business Venturing*, 18, 573–596.

Appleby, M. 2004. *Norfolk Arable Land Management Initiative (NALMI)*, Final Project Report June 1999 – May 2004, Countryside Agency.

Arbuckle, G., Lasley, P., Korsching, P. and Kast, C. 2009. *Iowa Farm and Rural Life Poll. Summary Report*. Iowa State University.

Baker, J. 2005. Why the obsession with succession? Minnesota Department of Agriculture, 90 West Plato Blvd., St. Paul, Minnesota.

Barclay, E., Foskey, R. and Reeve, I. 2005. *Farm succession and inheritance: comparing Australian and international research*. The Institute for Rural Futures, University of New England, Armidale, NSW, Australia.

Beddington, J. 2009. Speech to GovNet Sustainable Development UK 2009 conference, http://www.govnet.co.uk/news/govnet/professor-sir-john-beddingtons-speech-at-sduk-09 [accessed 28/05/09].

Berry, W. 2009. *Bringing it to the Table*. Berkley: Counterpoint.

Björnberg, Å. and Nicholson, N. 2008. *Emotional Ownership: The Critical Pathway Between the Next Generation and the Family Firm*. London: Institute for Family Business.

Boehlje, M.D. and Eidman, V.R. 1984. *Farm Management*. New York: John Wiley.

Bonnen, J. and Browne, W. 1989. Why is agricultural policy so difficult to reform? in *The Political Economy of US Agriculture: Challenges for the 1990s*, edited by C. Kramer. Washington: Resources for the Future, 7–33.

Brookfield, H. and Parsons, H. 2007. *Family Farms: Survival and Prospect*. Abingdon: Routledge.

Browne, W., Skees, J., Swanson, L., Thompson, P. and Unneverhr, L. 1992. *Sacred Cows and Hot Potatoes: Agrarian Myths in Agricultural Policy*. Boulder: Westview Press.

Bryden, J., Bell, C., MacKinnon, N. and Salant, P. 1992. *Farm Household Adjustment in Western Europe 1987–1991*. Final Report on structural change, pluriactivity and the use made of structures policies by farm households in the European Community. Nethy Bridge, Invernesshire: Arkleton Trust (Research) Ltd.

Burton, R., Mansfield, L., Schwarz, G., Brown, K. and Convery, I. 2005. *Social Capital in Hill Farming*. Aberdeen: The Macaulay Land Use Research Institute.
Burton, R. and Walford, N. 2005. Multiple succession and land division on family farms in the South East of England: a counterbalance to agricultural concentration? *Journal of Rural Studies*, 21(3), 335–347.
Calus, M. and Huylenbroeck, G. 2010. The persistence of family farming: a review of explanatory socio-economic and historical factors. *Journal of Comparative Family Studies*. HighBeam Research. 21 Jul. 2011 <http://www.highbeam.com
Calus, M., Huylenbroeck, G. and Lierde, D. 2008. The relationship between farm succession and farm assets on Belgian farms. *Sociologia Ruralis*, 48, 38–56.
Corsi, A. 2004. *Intra-family succession in Italian farms*. Les mutations de la famille agricole: Conséquences pour les politiques publiques, SFER Conference, Paris, 22–23 April 2004, available at http://www.child-centre.unito.it/ [accessed 21/07/11].
Danes, S., Stafford, K., Haynes, G. and Amarapurkar, S. 2009. Family capital of family firms: bridging human, social, and financial capital. *Family Business Review*, 22(3), 199–215.
Djurfeldt, G. 1996. Defining and operationalizing family farming from a sociological perspective. *Sociologia Ruralis*, 36(3), 340–351.
Dumas, C ., Richer, F. and Cyr, L. 1995. Factors that influence the next generation's decision to take over the family farm. *Family Business Review*, 82, 99–120.
Dyck, B., Mauws, M., Starke, F. and Mischke, G. 2002. Passing the baton: the importance of sequence, timing, technique and communication in executive succession. *Journal of Business Venturing*, 17, 143–162.
Errington, A. 1984. *Delegation on Farms*. Unpublished PhD thesis. The University of Reading: Department of Agricultural Economics and Management.
Errington, A. 1996. A comment on Djurfeldt's definition of family farming. *Sociologia Ruralis*, 36(3), 352–355.
Errington, A.J. and Lobley, M. 2002. *Handing over the reins: a comparative study of intergenerational farm transfers in England, France, Canada and the USA*. Agricultural Economics Society Conference, Aberystwyth, UK, 8–11 April 2002.
Fennell, R. 1987. *The Common Agricultural Policy of the European Community: Its Institutional and Administrative Organisation*. Oxford: BSP Proffessional Books.
Fischler, F. 2001. *International Aspects of European Agriculture*. Lecture at the Bologna Center of the Johns Hopkins University Bologna, October 30, 2001.
Frances, F. 1994. *Family Agriculture: Transition and Transformation*. London: Earthscan.
Galeski, B. and Wilkening, E. 1987. Conclusions, in *Family Farming in Europe and America*, edited by B. Galeski and E. Wilkening. Boulder: Westview Press.
Gasson, R. 1987. Family Farming in Britain, in *Family Farming in Europe and America*, edited by B. Galeski and E. Wilkening. Boulder: Westview Press.
Gasson, R. 1993. *The Nature of the Farm Family Business*. XV European Congress of Rural Sociology, Wageningen, Netherlands, August 1993.

Gasson, R. and Errington, A. 1993. *The Farm Family Business*. Wallingford: CAB International.

Gasson, R., Crow, G., Errington, A., Hutson, J., Marsden, T. and Winter, M. 1988. The farm as a family business: a review. *Journal of Agricultural Economics*, 39, 1–42.

Glauben, T., Tietje, H. and Weiss, C. 2004. Intergenerational succession in farm households: evidence from upper Austria. *Review of Economics of the Household*, 2, 443–461.

Godfray, C., Beddington, J., Crute, I., Haddad, L., Lawrence, D., Muir, J., Pretty, J., Robinson, S., Thomas, S. and Toulmin, C. 2010. Food security: the challenge of feeding 9 billion people. *Science*, 327, 812–818.

Gray, J. 2000. *At Home in the Hills: Sense of Place in the Scottish Borders*. New York: Berghahn Books.

Harl, N. 1972. The family corporation, in *Size, Structure, and Future of Farms*, edited by A. Ball and E. Heady. Ames: Iowa State University Press.

Hastings, M. 1984. Succession on Farms. Unpublished MSc thesis: Cranfield.

Hill, B. 1993. The 'myth' of the family farm: defining the family farm and assessing its importance in the European Community. *Journal of Rural Studies*, 9(4), 359–70.

Hoppe, R.A. and Banker, D.E. *Structure and Finances of U.S. Farms: Family Farm Report, 2010 Edition*, EIB-66, U.S. Dept. of Agr., Econ. Res. Serv. July 2010.

Keith, W. 1975. *The Rural Tradition: Non-Fiction Prose Writers of the English Countryside*. Sussex: Harvester.

Kennedy, L. 1999. Farm succession in modern Ireland: elements of a theory of inheritance, in *Rural Change in Ireland*, edited by J. Davis. Belfast: Institute of Irish Studies, The Queen's University of Belfast.

Laband, D. and Lentz, B. 1983. Occupational inheritance in agriculture. *American Journal of Agricultural Economics*, 65, 311–14.

Lobley, M. and Winter, M. 2009. Knowing the land, in *What is Land For? The Food, Fuel and Climate Change Debate*, edited by M. Winter and M. Lobley. London: Earthscan.

Lobley, M., Baker, J. and Whitehead, I. 2010. Farm succession and retirement: some international comparisons. *Journal of Agriculture, Food Systems, and Community Development*, 1(1), 49–64.

Marx, K. 1976. *Capital: A Critique of Political Economy*. London: Penguin.

Milton, L. 2008 *Unleashing the Relationship Power of Family Firms: Identity Confirmation*.

Mishra, A. and El-Osta, H. 2008. Effect of agricultural policy on succession decisions of farm households. *Review of Economics of the Household*, 6, 285–307.

National Assembly for Wales 2001. Farming for the Future. The Government of the National Assembly for Wales, Cardiff.

Northfield, Lord 1979. *Report of the Committee of Inquiry into the Acquisition and Occupancy of Agricultural Land*. Cmnd.7599. London: HMSO.

OECD 2010. *Agricultural Outlook 2010–2019*. Paris: OECD.
Paarlberg, D. 1980. *Farm and Food Policy*. Nebraska: University of Nebraska Press.
Parry, J., Barnes, H., Lindsey, R. and Taylor, R. 2005. *Farmers, Farm Workers and Work-related Stress*. London: Health and Safety Executive.
Pearson, A., Carr, J. and Shaw, J. 2008. Toward a theory of familiness: a social capital perspective. *Entrepreneurship Theory and Practice*, 32(6), 949–969.
Potter, C. and Lobley, M. 1992. Ageing and succession on family farms. *Sociologia Ruralis*, 32, 317–334.
Potter, C. and Lobley, M. 1996. Unbroken threads? Succession and its effects on family farms in Britain. *Sociologia Ruralis*, 36, 286–306.
Price, L. 2010. 'Doing it with men': feminist research practice and patriarchal inheritance practices in Welsh family farming. *Gender, Place and Culture*, 17(1), 81–97.
Ramos, G. 2005. The continuity of family agriculture and the succession system: the Basque case. *Journal of Comparative Family Studies*, 36(3), 367–377.
Rogoff, E. and Heck, R. 2003. Evolving resrecah in entrepreneurship and family business: recognizing family as the oxygen that feeds the fire of entrepreneurship. *Journal of Business Venturing*, 18, 559–566.
Rosenzweig, M. and Wolpin, K. 1985. Specific experience, household structure, and intergenerational transfers. Farm family land and labour arrangements in developing countries. *The Quarterly Journal of Economics*, 100, 961–987.
Sharma, P., Chrisman, J. and Chua, J. 2003. Predictors of satisfaction with the succession process in family firms. *Journal of Business Venturing*, 18, 667–687.
The Royal Institution of Chartered Surveyors 2011. *Rural Land Market Survey*. London: RICS.
Traill, B. 1979. An empirical model of the U.K. land market and the impact of price policy on land values and rents. *European Review of Agricultural Economics*, 6(2), 209–232.
Uchiyama, T., Lobley, M., Errington, A. and Yanagimura, S. 2008. Dimensions of intergenerational farm business transfers in Canada, England, the USA and Japan. *Japanese Journal of Rural Economics*, 10, 33–48.
United States Department of Agriculture National Agricultural Statistics Service (2010) Land Values and Cash Rents 2010. Available at: http://usda.mannlib.cornell.edu/usda/current/AgriLandVa/AgriLandVa-08-04-2010.pdf [accessed 21/07/11].
Whatmore, S., Munton, R., Little, J. and Marsden, T. 1987. Towards a typology of farm businesses in contemporary British agriculture. *Sociologia Ruralis*, 27(1), 21–37.
Winter, M. 1995. *Networks of Knowledge*. WWF-UK Agriculture Report.
Winter, M. and Lobley, M. 2006. *Family Farming in Flux: Reflections on Fifty Years of Family Farming Research in British Rural Sociology*. Agricultural History Society Winter Conference, London.
Zachary, R. 2011. The importance of the family system in family business. *Journal of Family Business Management*, 1(1), 26–36.

Chapter 2
Australian Farmers' Attitudes Toward Succession and Inheritance

Elaine Barclay, Ian Reeve and Roslyn Foskey

Introduction

Unlike other businesses, family farming is characterised by an intimate connection between the farm as a place of work, career and family tradition. What impacts upon one aspect, will impact upon all. An earlier study by the lead author (Kaine, Barclay and Stayner 1997) noted the difficulty and complexity of succession planning for Australian farmers as they sought to meet three conflicting objectives: to maintain a viable farm business for the next generation, treat all of their children fairly and provide for their own retirement. For some farmers, farming is a 'way of life' and they experience great difficulty in handing responsibility over entirely to their children. As one participant reported:

> The thought of handing over the farm to the next generation is a concept that is very hard to come to terms with. Means letting go of how I see myself as a landholder – and becoming a retiree – basically of no significance anymore. Therefore, I can't really approach the subject of transferral with the children. It's like the end of my life and I find it very difficult to face (Kaine et al. 1997).

As the vast proportion of Australian agricultural businesses are family owned and operated, how these families plan and manage retirement, succession and inheritance is therefore a concern for the whole agricultural industry. Effective advance planning in these areas can provide a sense of confidence and security and thus help preserve harmony within the family. However, if not properly handled, the transfer of the farm between generations can lead to confusion, uncertainty, suspicion and can result in deeply damaging divisions between family members.

These decisions have been made even more difficult by prolonged and severe drought and economic forces, which have left many of Australia's farm families unable to sustain operating budgets. Several farmers responding to the Kaine et al. (1997) study commented that 'today, leaving the farm to your children is a form of child abuse!' The difficulty in these decisions may lead to the issue being ignored until a critical event such as marriage, illness, or the death of a family member provokes action. The pressures that such events can create may not be conducive to calm and considered deliberations, nor effective communication between family members, resulting in decisions that have the potential to threaten the viability of the farm business. This chapter reports on the key findings of the International Farm Transfers Study conducted in Australia in 2004, which explored this issue further and compared Australian trends in career progression, succession, inheritance and retirement on farms with those in other countries participating in the International Farm Transfers Study.

The Farm Transfers Study in Australia

The Australian Farm Transfers study employed a literature review, a nationwide mail survey of farmers and statistical analysis of the data gathered. The survey instrument was refined for Australian conditions and piloted with 100 farmers before being mailed to 5000 farmers across the country in the winter of 2004. A reminder notice and questionnaire were sent to non-respondents after four weeks. The response rate to the survey after allowing for 'return to senders' was 36 per cent providing a sample of 1180 farm families for analysis.

The survey sample included 998 (85 per cent) males and 167 females (14 per cent). Their ages ranged between 17 and 94 years (Mean 54.45 years, SD 11.99 years). With the median age of farmers being 52 years (ABS 2008), the sample was representative of a cross-section of Australian farmers. The size of holdings ranged from a small hobby farm of one hectare through to a 1,381,800 hectare outback station in the Northern Territory. The most common types of agricultural production were beef, sheep meat and wool production, and cereal growing.

The majority of farm businesses were family partnerships, private or family companies, or sole operations. These findings reflect the structure of agriculture in Australia, which is mostly comprised of family farm operations although the number of farming families declined by 9 per cent between the 2001 and 2006 Australian Census of Population and Housing (ABS 2008). The average length of time respondents had been operating their property was 30 years (SD 13.69 years). Most farmers reported their farm businesses were in a good financial position which was a surprising finding considering the many years of severe drought experienced in Australia, particularly in the eastern states. However, this finding was associated with the number of smaller size properties within the sample. Nevertheless, many families are asset rich but cash poor, which can impact upon their retirement and succession plans. Most respondents (86 per cent) were full time farmers with at least one family member working on the property. However, 25 per cent also participated in off-farm work.

Retirement

Australian farmers' retirement plans were very similar to those of their counterparts in England. The larger proportion (56 per cent) planned to semi-retire at some stage. A sizable proportion (11 per cent) of respondents reported that they would never retire from farming. There were no significant differences in retirement plans between owners of small, medium or large properties. The international comparisons revealed that more Australian farmers planned to semi-retire than farmers in other countries. Foskey (2005) and Wilkinson (2009) term this 'retirement *in* farming.' As found in other countries, Australian farmers who had identified a successor tend to prefer semi-retirement. Uchiyama et al. (2008) surmised that the presence of a successor allows the farmer to reduce their level of involvement in the farm enterprise.

Australian farmers, on average, plan to retire or semi-retire at age 65. The majority of those who plan a full retirement intend to leave the farm between 65 and 69 years. Comparative analyses revealed that farm size was significantly associated with planned retirement age ($\chi^2 = 21.72$, $p<0.01$). Farmers on smaller farms were more likely to retire at 55 years or less while those on large farms were more likely to retire around 65 to 69 years. International comparisons revealed Australian farmers, along with their counterparts in Japan and the United States, intend to move into retirement at an older age than those in Canada, France and England.

Just under half (48.4 per cent) of the respondents planned to move from their current home when they retired or semi-retired. Of these, the most popular option reported by 57 per cent was to move to a nearby town. In Australia, this is a common strategy for the older generation in full or semi-retirement as it allows them to remain close to the property, be available if needed at busy times and continue to take an interest in the progress of the farming operation. Other respondents (18.6 per cent) expected to move to a smaller property or hobby farm upon retirement. Such a move enables the older generation to reduce the workload associated with farming but still maintain the lifestyle that they enjoy. These farmers are also more independent of the family farm that is now operated by the younger generation. Other farmers (4 per cent) envisaged moving to the coast. This is a phenomenon that is characteristic of Australian society as the coastal regions, particularly the north coast of New South Wales and southern coastline of Queensland, are popular destinations for retirees or for people seeking a 'sea change' (Hugo 1998; Foskey 2005). The survey also included the alternative of moving in with relatives. None of the Australian respondents chose this option, which reflects cultural norms in retirement preferences.

In comparison with other countries, Australian farmers' planned sources of retirement income are fairly evenly spread across a range of options. A slightly greater proportion of Australian respondents planned to support themselves in retirement through the sale of farmland and other farm assets compared with respondents in other countries. This finding may be a function of the larger number of smaller farms within the sample or it may reflect a trend amongst some Australian farmers to see farmland as superannuation. One respondent wrote:

> I have never intended to pass the farm to an heir. I always intended to develop the farm and realise a capital gain and use the profit from the sale after a unit has been purchased to assist with income (Farm Transfers Survey 2004).

As much of Australia's best farmland lies in the high population growth areas along the coast, selling land for a comfortable retirement is an attractive option for asset rich cash poor farmers (McAllister and Geno 2004). This finding also supports Voyce's, (1996) conjecture that an increasing number of farmers are taking a business approach to farming where farmland is merely a commodity that can be bought and sold. However, persistent drought and poor commodity prices in recent years have forced a number of farmers to sell their land and leave farming. It is estimated that up to half of the rural properties in south-eastern Australia could change ownership over the next decade (Minato, Curtis and Allan 2009).

The study also explored the extent to which respondents had discussed their retirement plans with family or sought professional advice. Previous studies (Kaine et al. 1997; Gamble et al. 1995; Crocket 2004; Gamble and Blunden 2004) have noted the difficulties farm families have in broaching such issues with family members. Kaine et al. (1997) found many farmers prefer to seek advice from accountants or solicitors rather than family members. The present study found that most respondents had talked to their spouse or partner, but less than half had discussed their retirement plans with other members of their family. Thus, over 50 per cent of younger generations in these families are being denied access to information that will significantly affect their future. On average, the older generation in these families is in their mid-fifties and the children are in their thirties and may have been working for at least ten years on the family farm. One respondent wrote:

> At 34 years old, my spouse and I find it very frustrating to be kept in the dark regarding succession. I have two sisters who will be considered (Farm Transfers Survey 2004).

However, compared with other countries, with the exception of Virginia USA, there is generally more discussion relating to succession and inheritance issues within Australian farm families. As in other countries, Australian farmers were more likely to have held family discussions if they had chosen a successor. This is a logical finding. However, more Australian farmers (39 per cent) discuss their plans with accountants than do their overseas counterparts. Twelve per cent of Australian respondents had not discussed these issues with anyone although this was a low proportion in comparison with farmers in other countries.

Succession

Just over half of the respondents (51.6 per cent) had identified a successor for their farm business. The proportion is comparable with those in England (52.8 per cent) and Japan (49.8 per cent), but greater than those in Ontario (39.8 per cent) or Quebec (42.1 per cent), and particularly Iowa (28.8 per cent) and Virginia (30.8 per cent) (Uchiyama et al. 2008). Notably, more of the younger farmers within the Australian cohort had chosen a successor in comparison to other countries.

Australian successors were most likely to be a son (52 per cent). Only 10 per cent were daughters. Although social changes have seen farm women increasingly take a more equal role in farm management (Farmar-Bowers 2009), it is evident in these findings that women are still viewed as dependents, either as a wife or daughter, and most continue to be excluded from inheritance of land. Most respondents noted daughters would inherit money or non-farm assets. Daughters were provided with a good education as another form of compensation. However, one respondent commented: 'More and more females are taking on the family farm and should be encouraged in all areas. Our girls are great.'

In the Australian study farmers tended not to consider succession before the age of fifty. Patterns in the relationship between age and identification of a successor were similar to patterns amongst English and Canadian farmers. However, Australian farmers have chosen a successor at a younger age in comparison to farmers in other countries. Farm size was a factor in these decisions as farmers on smaller farms were less likely to have nominated a successor. As Uchiyama et al. (2008) note, it is difficult to determine whether larger farms are more likely to have a successor in place or whether farmers who partner with a successor become larger. That is, whether it is the *succession effect* or the *successor effect* (Potter and Lobley 1996). However, in Australia, properties greater than 50,000 hectares were also less likely to have a successor. This is possibly because many large outback properties tend to be run as companies rather than simple family farm business structures.

The Farm Transfers Study is also interested in the degree to which successors take a *direct route* where they become involved in farming when they leave school or a *diversion route,* where they take up off-farm employment and return to the family farm at a later stage. In Australia, as in Canada and England, successors were most likely taking a *direct route* working alongside the older generation on the family farm. As found in most other countries, farm size clearly impacts upon successors' options as smaller properties are less able to support two generations. The study found successors on smaller farms were more likely to be working off-farm while those on large farms were more likely to be working full-time on the family farm ($\chi^2 = 20.51$, $p<0.01$).

Other factors also play a significant role in succession plans. For example, a recent study which considered the implications of drought and water scarcity for farm succession within the Goulburn Murray Irrigation District in Victoria, Australia (Bjornlund and Rossini 2010) found that in 2008 half of the farm households surveyed in the district were obtaining between 75–100 per cent of their household income through off-farm sources. Furthermore only 20 per cent of irrigators expressed any certainty of farm continuity, highlighting the impact climactic conditions can have on farm succession.

Delegation of Managerial Responsibilities

One of the main focuses of the International Farm Transfers Project was to determine the extent to which successors are involved in the management of the family farm. On Australian farms, as is the case on farms in all other countries within the study, financial decisions are the last responsibility transferred to the younger generation. The main differences in the Australian data were that successors on Australian farms were less likely to have control over decisions regarding the long-term balance and type of enterprises on farm or decisions regarding the purchase machinery or equipment. This may possibly be due to the larger size of operations in Australia and the size and cost of machinery required. Across all tasks, Australian farmers, while considerably more generous in delegating responsibility to successors than their counterparts in Iowa, are less generous than farmers in Canada and France. As Errington and Lobley (2002) note there may be a variety of policy, economic and social reasons for these differences that need more extensive investigation. For example, like English farmers, Australian farmers are more likely than those in the other countries to rely on farm income in retirement and consequently have an interest in protecting their future security by maintaining managerial control (Errington and Lobley 2002).

The final stage of the analysis assessed patterns in succession as defined by Errington (1998) according to the degree of partnership between the older and younger generation in managing the family property. Of particular interest was the *farmer's boy* situation (Errington 1998), where the successor works on the family farm for many years but is allowed few responsibilities. Successors in this position have little opportunity to develop managerial skills necessary to effectively manage a farm business. This situation has been found to be more common in England than in other countries (Errington and Lobley 2002). There was a relatively small proportion of Australian successors that were in this category (5.3 per cent). One of these successors wrote:

> The biggest problem is the 'old bloke' not handing over the reins – I'm now in my late 40s and the plan has now changed from just me inheriting the farm (my sisters will inherit share portfolios) to provision for nieces and nephews all getting land or proceeds from the sale (Farm Transfers Survey 2004).

One farmer acknowledged the position his son was in:

> The farm will pass to the son currently running farm who has worked for virtually no wages for 16 years; other children have incomes from occupations separate from the farm. The farm is barely viable for one family; if left to more than one child; it would have to be sold. The son is to pay five siblings a set amount of cash, which was arrived at by considering the years of unpaid wages being offset by building up his equity in the farm (Farm Transfers Survey 2004).

Another successor option was the *separate enterprise* where successors were working on the farm but held full responsibility for a particular enterprise within the farm business; in most cases these were cropping or livestock enterprises. Other successors (24 per cent) were in the best position of having a full partnership with the older generation. Most of this group also held full responsibility for a separate enterprise within the farm business. Although most successors were working on the family farm, more Australian successors were taking a *professional detour*, either running a non-farm business, employed on another farm or working in off-farm employment than successors in other countries (with the exception of successors in Iowa). Twelve per cent had a *standby holding,* where they were running their own property. This percentage was greater than other countries with the exception of Iowa.

Factors Impacting upon Succession

To identify those factors that are the greatest predictors of farmers choosing a successor, a logistic regression was conducted. The analysis revealed that the factor that had the greatest power in predicting whether farmers had chosen a successor was their adherence to particular cultural mores about succession and inheritance (see Table 2.1). This finding was irrespective of age or education of the respondents. Farmers who desired to retain the family farm as a whole unit and pass it onto one heir were more likely to have chosen a successor. These cultural mores largely emanate from predominantly Anglo Saxon traditional approaches to succession and inheritance that uphold values of patriarchy and primogeniture. Other predictors of choosing a successor were a larger property size and a greater length of time the farm family had been operating the property and whether or not the respondents planned to retire or semi-retire.

The influence of cultural mores upon succession planning was assessed by a question that asked respondents what they believed was the best plan for the succession of their family farm operation and why. Their responses were collapsed into categories as defined by the Farm Transfers Study. These included passing the farm to (1) a sole successor; or (2) to all those with inheritance rights within the family, or (3) to those living together within the farm family. Most respondents believed that passing the farm onto a sole heir was the best way to maintain the farm within the family. As one stated:

Table 2.1 Logistic regression coefficients of factors significantly predicting upon farmers' decisions to choose a successor

Variable	Beta Scores	Standard Error
Belief farm should be passed to sole heir	1.1823***	.2385
Alternative inheritance plans	-2.6343***	.4045
Length of time on property	-.6005***	.1083
Farm size	.3894*	.0186
Constant	.8859*	.4342
-2 Log-Likelihood	202.95***	

Note: *p<0.01 **p<0.001 ***p<0.0005 (two tailed tests)

> Passing the farm onto one heir guarantees the best chance of the farm surviving and staying in the family. Giving out assets in property to non-participating family members is destroying the family farm today. The one who inherits the farm should be actively working the land; too many now get left farming land ownership that have never done any work on the property (Farm Transfers Survey 2004).

Others believed that the farm should be left to the child/children who displayed the most interest in farming: 'My farm will be left to my children, according to their commitment, interest and input to this operation.' Others noted that the need to ensure other family members receive an equal share of the inheritance could cause problems for the successor.

> It is always difficult for the one/s who remain on-farm. They battle everything, with often little financial reward for time and effort, while the one/s who leave and want their share end up taking any of the profits leaving the one/s remaining in a situation of being unviable, and in turn nearly unemployed, with no remaining capital value (Farm Transfers Survey 2004).

Some reported they would sell the property to ensure that all their children would receive an equal share in the inheritance. The farmers in the best position had a property set aside for each child and sufficient assets to support their retirement. Others who found the decisions too difficult had decided to let their children work it out for themselves after they had gone. Some intended to sell the family property because there was no successor or because their children were not interested in farming, or because they needed funds to support their retirement.

Two-thirds of the respondents were farming land that had been in their family or their spouse's family for several generations. Most farms had been family holdings for at least three generations while a few had been held in a family for six or seven generations. These respondents expressed pride on the length of time their property had been in their family. The remaining third had purchased their property and had no family connection to the land. These findings confirm Nalson and Craig's (1987) statement that despite common perceptions of rural Australia, generations of farm families on particular tracts of land is far from the norm. In reality, what is passed on to the younger generation can be the attachment to the occupation of farming.

Those who reported a long history of familial connection to their land were mostly of English heritage. There was no significant relationship found between cultural heritage and attitudes towards succession and inheritance. However, aspects of the relationship were consistent with Salamon's (1984, 1985, 1987) findings in studies of American farmers' heritage where those of German heritage preferred that all family members share in inheritance. In the present study it was interesting to note that a group of respondents who were descendents from a diverse range of ethnic backgrounds, such as Russian, Hungarian and Basque, displayed a clear preference for ensuring all family members have an equal share of the farm business (see Figure 2.1). This finding suggests that it would be of interest to extend the International Farm Transfers Study to these countries. For example, Missingham, Dibden and Cocklin (2006) suggest that social and economic inequality in rural Australia is an important influence on the succession plans of people from non-English speaking backgrounds (NESB), with a trend for the senior NESB generation on small farms to invest in education and social mobility for their children, leading to the second generation leaving the agricultural sector.

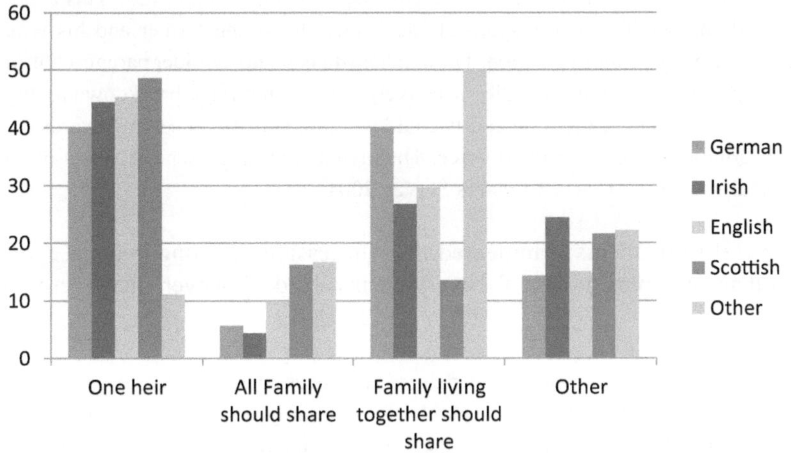

Figure 2.1 Attitudes to the division of property by cultural heritage

The study also revealed several other factors that impact upon farmers' plans for succession and inheritance. These included the possibility of divorce within the family and the subsequent loss of all or part of the family farm in Family Court settlements, the impact of assets and income tests on eligibility for aged pensions, the impact of government taxes, the financial pressure caused by escalating costs and the persistent drought. The need to preserve the viability of a farm business was an important factor determining farmer's decisions to sell or pass the farm on. As one stated: 'If it's not debt-free, don't hand it down.' Another commented:

> At my age, I intend to work towards (and with my sons) achieving viable land units to enable those wanting to stay on the land to do so. My father helped me buy my first property and is a great source of advice and support. I would like to do this for my sons as well (Farm Transfers Survey 2004).

Several respondents noted the impact of divorce upon farm family partnerships and the succession process. One wrote:

> In our area, farm inheritance is a hot issue. A number of properties have been sold up when marriages break up and the daughter-in-law is awarded half of a fourth-generation family property. This has often meant that the parents delay passing over ownership to the working sons on the property. Consequently, middle-aged farmers are disgruntled, and rightly so, as they have no controlling powers: they are told by their aging parents, 'It's yours when I die.' This is not good enough. To make responsible decisions, the young farmer and his wife need to be in a control position. The antagonism is set up, as older parents control the purse strings until they die, is as likely to cause marriage breakdown as any other situation. Perhaps the courts could be more realistic in setting awards for disgruntled wives after the divorce. The high awards at present being set often break the farmer (Farm Transfers Survey 2004).

Several respondents commented that succession planning was an extremely difficult and complex process for everyone involved. However, there were success stories:

> In a very complex problem, our parents have been lucky enough to have: (a) a large enough area to divide into viable areas; (b) only two sons, one who wanted to go straight onto the land, the other had other interests; (c) adequate off-farm investments to support the parents and not a high level of debt; and (d) finally, an absolute belief that the sons must be masters of their own destinies without parental interference! (Farm Transfers Survey 2004).

Discussion

One consistent theme within these findings is the persistence and fervent adherence to a rural ideology that strongly influences farm succession planning within Australian farm families. This is not a new finding. The role of agrarian ideology in family farming has often been the subject of previous research (e.g. Gray 1991; Craig and Phillips 1983). However, there is also evidence of social change in rural Australia with changing roles for women, more children taking up tertiary education and greater diversity in farm business structures (Farmar-Bowers and Lane 2009; Farmar-Bowers 2009; Mendham and Curtis 2007). While the present study found evidence of these social changes, it also found that traditional norms and values still have a disproportionate influence on farm families' decisions regarding the management of the farm business. It appears, therefore, that succession and inheritance are likely to remain difficult and complex processes for many farmers.

Most respondents believed that passing the farm onto a sole heir was the best way to maintain the farm within the family. These findings support Goodman and Redclift's (1981) conjecture that primogeniture has been the dominant pattern of inheritance since European settlement in Australia. These practices are supported by an *absolutist* concept of land ownership, which implies that everything has an owner and every owner has total control over the things he or she owns (Reeve 2002). This is accompanied by a belief in the family farm as the appropriate production unit, the legitimacy of family ownership by application of their labour, and in farmers being the backbone of Australia's economy and society (Berry 1990, cited in McAllister and Geno 2004). Therefore, norms and laws regarding inheritance are a public and social response to private property rights. While Voyce (1994) maintains that more Australian farmers are taking a business approach to farming and land ownership, the present study suggests that traditional values are pervasive within rural Australia. Salamon (1984, 1985 and 1987) claims that a family's perception of its relationship to land has profound influences on its operational style including their approach to succession and inheritance.

The study also revealed that family partnerships or sole proprietorships are the primary types of farm family legal structures within Australia. McAllister and Geno (2004) suggest that traditional legal structures of property ownership are used by farmers as legitimate means of protecting values pertaining to property, family and inheritance. However, the authors did find that younger farmers favoured the newer forms of business structure. There were similar findings in the present study. Thus, younger farmers may be questioning the relevance of traditional legal structures and the future may see farmland being sold in retirement or alternatively succession and inheritance managed through family trusts or company structures. The study also found that only a few farms had been in the family longer than three generations and that the majority of respondents had purchased the land they farmed. Thus, while the ideals persist, in reality, it is not the case.

The study found that rural ideology also impacts upon attitudes towards retirement. The status of farming as an occupation where individualism, hard work and utilitarianism are highly valued means that retirement is not well regarded. In these circumstances, and in particular for farmers without off-farm interests or hobbies, the only reason for slowing down is decreased physical capacity to work, meaning that rather than retirement being a planned transition it risks being a negative experience, forced upon the farmer through ill-health (Foskey 2005; Wilkinson 2009). The present study found that most Australian farmers prefer semi-retirement. Most intended to move to town, which allows for continued involvement at various levels in the farm business. In comparison with other countries, Australian farmers' intended sources of retirement income are fairly evenly spread across a range of options although a slightly greater number of respondents planned to support themselves in retirement through the sale of farm land and other farm assets. However, some of these farmers may not be able to realise this goal if they are unable to sell.

It was evident in the respondent's comments that factors of viability and low farm incomes mediate in the relationship between ideology and succession planning. The past two decades of drought, low commodity prices and changes in land management has seen a downward trend in real incomes of farmers (Bjornlund and Rossini 2010; Gray and Lawrence 2001; Hicks, Basu and Sappey 2008). Thus, while the majority of respondents had a good debt/asset ratio, those who are asset rich and cash poor face obstacles in succession and inheritance planning. This is reflected in the trend for the older generation to insist that the younger generation receives a good education to give them options in career choice. While they cling to ideals of passing the farm within the family, reality has led farmers to seek other alternatives for their children.

The influence of Anglo Saxon traditions towards farming was also reflected in the many similarities between Australian farmers and their English counterparts in attitudes and patterns in transfer of managerial responsibility to successors, which occurred at a much slower rate than in other countries. However, this slower progression of succession may also be a consequence of more retiring farmers depending upon the farm as income in retirement. In such circumstances, the older generation may never relinquish legal control often retaining ownership of the land until death and therefore maintaining a measure of control over the farm business. This will impact upon the succession process and the rate of transfer of managerial control to successors.

As previous studies of farm succession in Australia have found, the present study revealed that issues surrounding retirement, succession and inheritance tend not to be discussed with family members. While it cannot be concluded that all these families will face difficulties managing the succession process, the lack of communication increases the likelihood that problems will arise as plans are made on the basis of misunderstandings and mistaken expectations. In addition, the longer discussions are delayed, the fewer options and opportunities are available for the family to take remedial action (Kaine et al. 1997). However, the fact that so many respondents were concerned about farm viability suggests that family discussions may be hampered by the current economic climate in rural

Australia. Discussions may be delayed until prospects look more promising. Many respondents were concerned about passing the farm on when children could earn significantly more in non-farming occupations.

It can be concluded that rural ideology significantly impacts upon Australian farmers' attitudes and values, and consequently the way in which they approach retirement, succession and inheritance. However, it is evident that younger farmers are questioning these traditional approaches, as economic and social changes in recent times have revealed the unsustainability of current practices. Therefore, there is opportunity for change. Nevertheless, future policies and programmes relating to succession and inheritance in Australia will have to take into account the pervasiveness of rural ideology and work within this structure to effect change.

Acknowledgment

The research that forms the subject of this chapter has been supported by a grant from the Rural Industries Research and Development Corporation, Canberra, ACT, Australia.

References

ABARE (Australian Bureau of Resource Economics) 2001. *Australian Farm Surveys Report 2000*. Canberra, ACT: Australian Bureau of Resource Economics.

ABS (Australian Bureau of Statistics) 2008. *Agriculture in Focus: Farming Families, Australia 2006*, Publication No. 7104.0.55.001. Canberra, ACT: Australian Bureau of Statistics [online]. Available at: http://www.abs.gov.au/AUSSTATS/abs@.nsf/mf/7104.0.55.001/ [accessed 20th February 2012].

Bjornlund, H. and Rossini, P. 2010. *Climate Change, Water Scarcity and Water Markets – Implications for Farmers' Wealth and Farm Succession*, paper presented to 16th Pacific Rim Real Estate Society Conference, Wellington, New Zealand, January 2010.

Craig, R.A. and Phillips, K. 1983. Agrarian ideology in Australia and the United States. *Rural Sociology*, 48(3), 409–420.

Crocket, J. 2004. The nature of farm succession in three New South Wales communities. *AFBM Journal*, 1(1), 14–27.

Errington, A.J. 1998. The inter-generational transfer of managerial control in the farm-family business: a comparative study in England, France and Canada. *Journal of Agricultural Education and Extension*, 5, 123–136.

Errington, A.J. and Lobley, M. 2002. *Handing over the reins: a comparative study of intergenerational farm transfers in England, France, Canada and USA*, paper presented to The Agricultural Economics Society Annual Conference, Aberystwyth, NSW, 8–11 April 2002.

Farmar-Bowers, Q. 2009. Understanding the strategic decisions women make in farming families. *Journal of Rural Studies*, 26(2), 141–151.

Farmar-Bowers, Q. and Lane, R. 2009. Understanding farmers' strategic decision-making processes and the implications for biodiversity conservation policy. *Journal of Environmental Management*, 90(2), 1135–1144.

Foskey, R. 2005. *Older Farmers and Retirement*, report to the Rural Industries Research and Development Corporation, Canberra, ACT.

Gamble, D. and Blunden, S. 2004. Talking agribusiness: planning farm succession as a family, *Ground Cover*, 49 [online]. Available at: http://www.grdc.com.au/director/events/groundcover?item_id=publication-issue50&article_id=482B58E3ADF7859A1F775422578B90F3 [accessed 20 February 2012].

Gamble, D., Blunden, S., Kuhn-White, L. and Voyce, M. 1995. *Transfer of the Family Farm Business in a Changing Rural Society*, research paper No. 95/8. Canberra, ACT: Rural Industries Research and Development Corporation.

Goodman, D. and Redclift, M. 1981. *From Peasant to Proletarian: Capital Development and Agrarian Transition*. Oxford: Blackwell.

Gray, I. 1991. Family farming and ideology: some preliminary exploration, in *Family Farming, Key Papers No. 2*, edited by M. Alston. Wagga Wagga, NSW: Centre for Rural Social Research, Charles Sturt University, 52–65.

Gray, I. and Lawrence, G. 2001. *A Future for Regional Australia: Escaping Global Misfortunes*. Melbourne, VIC: Cambridge University Press, Cambridge Institute of Agricultural Science.

Hicks, J., Basu, P.K. and Sappy, R.B. 2008. 55+ and working in an established rural regional Australian labour market. *Employment Relations Record*, 8(1), 1–16.

Kaine, G.W., Crosby, E.M. and Stayner, R.A. 1997. *Succession and Inheritance on Australian Family Farms*, TRDC Publication No. 198. Armidale, NSW: The Rural Development Centre, University of New England.

McAllister, J. and Geno, B. 2004. Australian farm inheritance-new patterns of legal structure in property rights and landholding. *Rural Society*, 14(2), 178–191.

Mendham, E., and Curtis, A. 2010. Taking over the reins: trends and impacts of changes in rural property ownership. *Society and Natural Resources*, 23(7), 653–668.

Minato, W., Curtis, A. and Allan, C. 2009. *Social Research looking at NRM investment and Demographic Change*, paper presented to 18th World IMAS/MODSIM Congress, Cairns, Australia, 13–17 July 2009 [online]. Available at: http://www.mssanz.org.au/modsim09/F12/minato.pdf [accessed 20 February 2012].

Missingham, B., Dibden, J. and Cocklin, C. 2006. A multicultural countryside? Ethnic minorities in rural Australia. *Rural Society*, 16(2), 131–150.

Potter, C. and Lobley, M. 1996. Unbroken threads? Succession and its effects on family farms in Britain. *Sociologia Ruralis*, 36, 286–306.

Reeve, I. 2002. *Property Rights and Natural Resource Management*, Institute for Rural Futures Occasional Paper 2002/1. Armidale, NSW: University of New England.

Salamon, S. 1984. Ethnic origin as explanation for local land ownership patterns, in *Research in Rural Sociology and Development: Focus on Agriculture*, edited by H. Schwarzweller. AI Press Ltd, 161–186.

Salamon, S. 1985. Ethnic communities and the structure of agriculture. *Rural Sociology*, 50, 323–340.

Salamon, S. 1987. Ethnic determinants of farm community character, in *Farm Work and Fieldwork: Anthropological Perspectives on American Agriculture*, edited by M. Chibnik. Ithaca, NY: Cornell University Press, 16–188.

Uchiyama, T., Lobley, M., Errington, A. and Yanagimura, S. 2008. Dimensions of intergenerational farm business transfers in Canada, England, USA and Japan. *Japanese Journal of Rural Economics*, 79(5), 33–48.

Voyce, M. 1994. Testamentary freedom, patriarchy and inheritance of the family farm in Australia. *Sociologia Ruralis*, 34(1), 71–83.

Wilkinson, R.L. 2009. *Population Dynamics and Succession Strategies of Rural Industry Producers*, thesis presented to fulfill the requirements of the Degree of Doctor of Philosophy, Victoria University, Australia [online]. Available at: http://eprints.vu.edu.au/1943/ [accessed 20 February 2012].

Chapter 3
New Patterns of Succession in the Australian Wool Industry

Roger Wilkinson

Introduction

Woolgrowing in Australia has a long history. The industry has not only a long tradition of inheritance of both the family farm and the occupation of farming, but also a widespread personal identity based around being a woolgrower that has been maintained and reinforced often over several generations (Anderson 1966). Both the tradition of inheritance and the identity developed around woolgrowing now appear to be under threat.

The 1990s were not kind to woolgrowers in Australia. Wool prices throughout the decade remained at historically low levels, resulting in low farm incomes (Martin 1998: 13). Woolgrowers responded to the tough times by diversifying into other agricultural industries such as prime lambs, cropping and even bluegum trees where they could. However, in the specialist woolgrowing areas, alternatives to wool are few, so diversification was often not a realistic option. Many woolgrowers were able to maintain their incomes only through off-farm work.

These pressures that were specific to the wool industry added to generic, long-term, pressures on rural youth all over the world to seek greater economic and social opportunities in the city (e.g. Bowsfield 1914, Jackson-Smith and Barham 2000). The result was a decreased likelihood of inter-generational transfer of woolgrowing farms in some areas (e.g. Curtis et al. 2000). Even when succession did occur, new patterns of entry to and exit from woolgrowing emerged as succession arrangements became more diverse. In this chapter the recent changes in succession arrangements in the Australian wool industry are explained.

Late Succession and Intergenerational Adjustment

Several American studies have observed a sharply reduced entry of young people into farming (Smith 1987, Hoppe 1996, Buttel et al. 1999, Gale 2000). This is not solely a recent phenomenon (Bowsfield 1914: 9–10, Kanel 1961). Tolley and Hjort (1963: 32) observed that, 'Adjustment comes primarily through reduction of entry of youths into farming.' Reduction in entries to farming may be caused by entry barriers (such as lack of capital or knowledge) or occupational choice (potential entrants decide there are better opportunities elsewhere). In the USA, reduced entry has occurred despite government programmes that attempt to overcome entry barriers, so it is likely that occupational choice is the major contributor (Gale 2003). The traditional pool of replacement farmers, young people raised on farms, has declined, partly due to off-farm migration of farm children, but also because of a declining number of children born to farm women (Hoppe 1996).

Much of the reduction in farmer numbers observed in several Australian studies in the 1970s also appears to have come from the reduced entry of young farmers into the industry (Hawkins and Watson 1972, Salmon, Fountain and Hawkins 1973, Bell and Nalson 1974). In one large-scale Australian study, 90 per cent of farmers said they intended to continue full-time work on the farm, yet only 60 per cent considered that any of their dependents would take over the property full-time (Salmon, Fountain and Hawkins 1973).

Where farms are small and viability is marginal, farming parents appear to deliberately equip their children with the education required for them to obtain remunerative city-based employment. In a study of an area in south west Victoria of predominantly small farms with low equity that had been taken up by soldier settlers about 20 years previously, 70 per cent of the children of respondents had left the district, mostly to go into 'white collar' jobs (Hawkins and Watson 1972). Education levels of the children who had left school were much higher than their parents. The level of education desired by parents for their children who were still at school was even higher still, and most of these parents desired non-rural employment for their children (Hawkins and Watson 1972). Napier (1972), who studied woolgrowers in the New England region of New South Wales at a time of low wool prices, noted that parental insistence on the son's obtaining training for an off-farm career in difficult times for farming may well delay the son's return to the farm or, perhaps even result in the son not returning at all.

Succession often occurs late. Errington and Tranter (1991) found that the average English successor was never allowed to decide when to pay bills during his father's lifetime. Succession late in the life of the successor can occur for two main reasons. One is that the father does not want to hand over control to the successor, perhaps because of a fear that the successor's marriage may break up (Price and Evans 2006). When the successor marries, the character of the farm family changes from 'nuclear' to 'extended', with two generations of families. This confounds the singular relationship between family and farm and it is during this period that the relation between family and farm is in greatest danger. The successor's marriage is two-edged. On the one hand, it is the basis for forming the next generation and thus necessary for keeping the farm in the family. On the other hand, the daughter-in-law becomes a claimant on the farm's assets and represents a threat to the future of the farm-family relationship if the son's marriage fails (Gray 1998). In Weigel and Weigel's (1987) study, daughters-in-law reported the highest frequency of stressors of all family members, and were clearly vulnerable. Parental paranoia about their daughter-in-law divorcing their son may be a self-fulfilling prophecy (McGuckian et al. 1995).

The other reason for late succession is that the farm is too small to support both generations, so the successor stays off the farm, perhaps helping out occasionally, waiting until the death or incapacity of the father before taking over the farm, perhaps part-time. In Potter and Lobley's (1996) study, these farms (where a successor was identified but was not present on the farm) were the most stable and least dynamic. Lacking both the incentive and ability to expand operations, but needing only to keep options open for the time the successor eventually returns, these farms were the least likely to have invested in major development or bought or sold land over recent years.

While the first form of late succession can occur on any size farm, the second form generally occurs on small farms that are often economically marginal. When the successor eventually takes ownership and control of the farm at the moment of inheritance on the death of the farmer, they are unlikely to be able to earn a full-time living from the farm. If the farm has been run down while the successor was not present, it may take more money or time to put right than the successor is willing to invest. The result is likely to be an even more pluriactive small farm sector (Potter and Lobley 1996).

Another way around the transition problem is for the succeeding child to take a job off the farm for a time, perhaps even until the parents give up farming. This 'occupational by-pass', as Blanc and Perrier-Cornet (1993) called it, requires both occupational and locational mobility on the part of the child, who needs both the education and skills required to obtain and succeed in the other job, and the willingness to work (and perhaps even to live) elsewhere. There is a risk that the child may become settled in the other job or place (Blanc and Perrier-Cornet 1993).

The rest of the chapter is an exploration of these issues in the context of woolgrowers in Australia. First some quantitative analysis of demographic trends is presented followed by the findings from a qualitative study of woolgrowing families who have recently faced or are considering succession issues. Both studies are described in more detail in Wilkinson (2009).

Demographics of Succession in Sheep Farming

The Census of Population and Housing (CPH), conducted every five years by the Australian Bureau of Statistics (ABS) allows sheep farmers to be identified and enumerated. This section contains a brief analysis of the demographic trends in the Australian sheep farming industry from 1976 to 2001, using the analytical method developed by Barr (2004). Although the focus of this chapter is woolgrowers, the census does not split sheep farmers into those who run sheep for wool and those who run sheep for meat. Most specialist sheep producers are, however, likely to be woolgrowers, since 86 per cent of Australia's adult sheep flock is made up of merino sheep, whose main produce is wool (Curtis and Croker 2005).

The number of sheep farmers in Australia reduced by more than half from 1976 to 2001 (Figure 3.1). This decline in numbers was not uniform across the period, with the sharpest falls over the periods 1976–1981 and 1986–1991. The decline in numbers was most pronounced among the younger age groups (Figure 3.2).

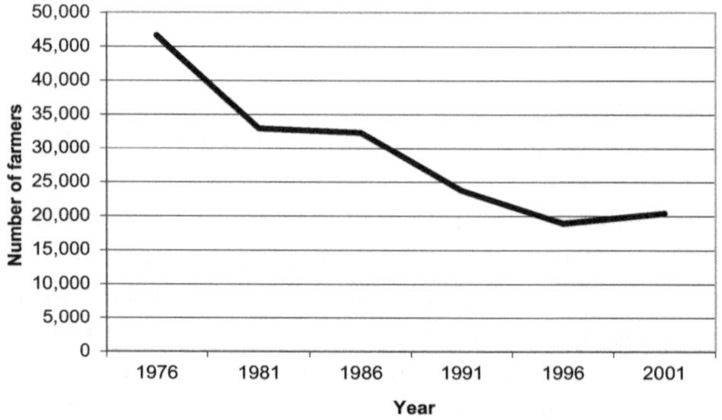

Figure 3.1 Number of Australian sheep farmers, 1976–2001
Source: derived from ABS CPH data

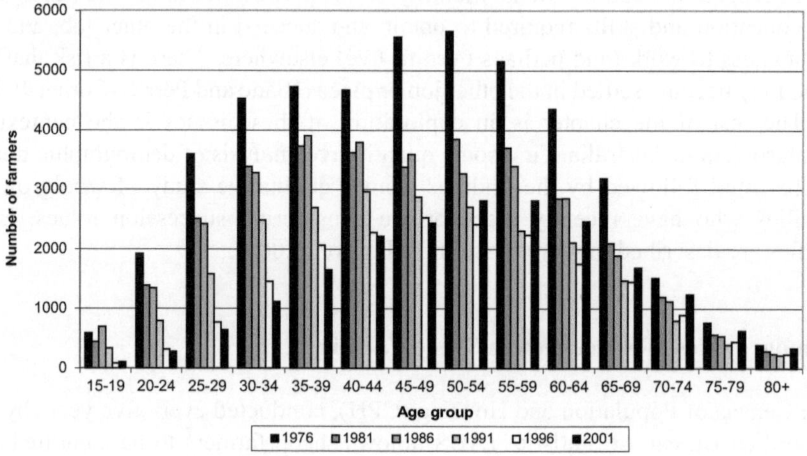

Figure 3.2 Number of Australian sheep farmers by age, 1976–2001
Source: derived from ABS CPH data

The exit rate has been falling since its retirement-driven peak in 1986–1991 during the period of relatively high wool prices (Figure 3.3). The entry rate has, however, remained relatively constant. Where the exit rate exceeds the entry rate, total numbers decrease. What has changed about entry to sheep farming has been the median age of entrants. The substantial drop in the number of young sheep farmers, as shown in Figure 3.2, has been driven by a large reduction in the number of young people entering sheep farming, particularly those aged under 35 (Figure 3.4).

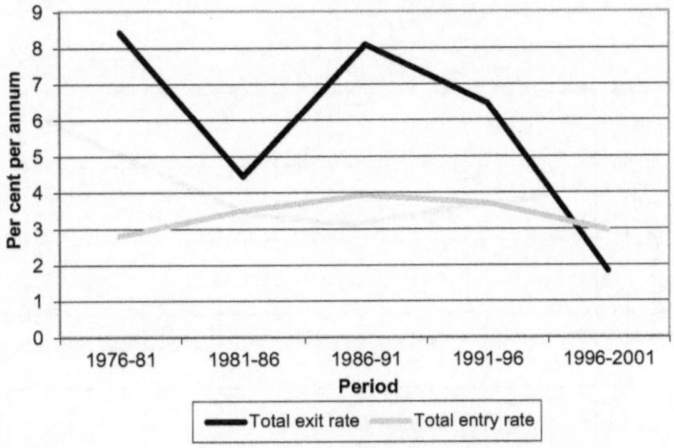

Figure 3.3 Entry and exit rates for Australian sheep farmers, 1981–2001
Source: derived from ABS CPH data

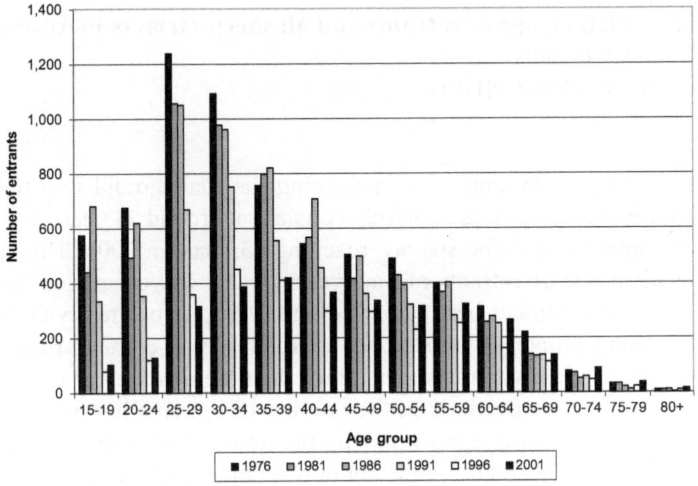

Figure 3.4 Number of entrants to Australian sheep farming by age, 1981–2001
Source: derived from ABS CPH data

The consequence of these trends is that the median age of entrants to sheep farming, relatively constant from 1976 to 1991 at between 33 and 35 years, increased by over seven years to 43 years in the 10-year period from 1991 to 2001 (Figure 3.5). This has in turn driven up the median age of all sheep farmers over the same period.

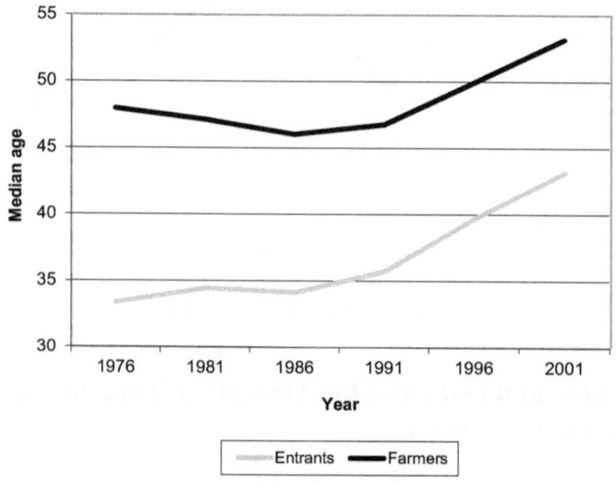

Figure 3.5 Median age of entrants and all sheep farmers in Australia, 1976–2001

Source: derived from ABS CPH data

The median age of entrants to sheep farming has gone through two main phases since 1976 (Figure 3.5). It was relatively constant at around 34 years from 1976 to 1986, then commenced rising steeply, reaching 43 years in 2001. This represents a rise in median age of entrants of nine years over a 15-year period. During the period of rapid adjustment between 1986 and 1991, associated with high wool prices and strong property turnover, the median age of entrants rose little, but during the period of slower adjustment since then the median age of entrants has risen rapidly. While the decrease in the number of entrants has now slowed, the increase in their median age has not. The previously common younger entry is now relatively rare. Sheep farming is not now a young person's game.

Archetypal Woolgrowers

To understand in more detail the changing nature of succession in the wool industry, a series of semi-structured interviews was conducted with 23 woolgrowing families spread throughout the Australian state of Victoria. The names of potential interviewee families were obtained from local informants, who were asked to suggest a variety of potential interviewees for whom entry, exit or succession would be likely to be a salient issue. Interviewees were selected to provide a broad spread of family circumstances, life stage, farm size, district and succession experiences.

Several archetypes among the interviewees were identified, farmers who exemplified particular characteristics of woolgrowers and succession patterns among woolgrowing families. These have been labelled and written up as brief character sketches. Each is based on one interview. To maintain continuity, some of the quotations have been edited slightly. Within the quotations, the interviewer's questions are in square brackets.

Rusted-on Farmer

Arthur's family has been in the district for nearly 130 years and on the present farm for 80 years. His father was passionate about sheep and wool, and so is Arthur. Several times in our interview I had to stop him explaining some special attribute he had bred into his sheep or how good their wool was and bring him back to the topic I wanted to discuss. 'I just like growing wool. I like good wool sheep. You just like the look of the sheep.' Arthur's wife recognises his enthusiasm.

> He's very proud of his sheep . He's been very good at it. Everything he does, he does well. It's a real challenge to grow merinos. Anybody can go and buy a mob of sheep and buy a fat lamb sire and put them over, but to breed merinos, where you've got to get rid of all the faults and breed the ideal animal, which is very difficult, and he's been trying that ever since he started growing sheep.

He talks about nothing but sheep, so much so that his wife had to learn to talk about sheep too. 'I'm a qualified woolclasser. It was either learn to talk sheep or have nothing to talk about. Because they weren't interested in how many rows did you knit today. So, in self-preservation I learnt, I talked about wool.'

Arthur left school and came home on the farm at 15 because his father was ill and could not run the farm. This was during the Second World War, and labour was short. He has been on the farm ever since, and is now in his mid-70s. The farm is managed very conservatively, understocked.

Arthur and his wife have several children. All but one has left the farm for other jobs. The remaining son, who works on the farm, is unmarried. His parents say they hope he has children of his own, to ensure the succession of the farm. Meanwhile, they are watching their other children's children for signs of interest in farming. Arthur's father left the farm to Arthur's children, in a complicated structure. This makes succession planning difficult, but Arthur and his wife think they will be able to make it all work out.

Arthur is not interested in any kind of retirement, despite recognising that he is less and less able to work as hard as he used to. His wife is keen to try other things in life, but Arthur is unable to focus on that and she knows it. [Do you want to retire?]

Husband:

No, I don't want to retire. I'm quite happy to keep on going around the place.

Wife:

I could retire and go wandering round Australia.

Husband:

We've been around, and we can still go around … I use a 4-wheel bike to get round the place. That's far easier than driving a vehicle, which you've got to climb in and out of … A fence got knocked over this morning. I just put some steel posts and wire on the back on the 4-wheel bike and go and fix it up.

[So that makes it easier to keep running the farm as you're getting older.]
Husband:

Oh, yes. [But there'll come a time when you can't. Do you want to keep on going until you physically cannot do it anymore, or do you want to actually stop trying to farm day-to-day and either go live somewhere else or let your son do the whole lot and you just give him a bit of a hand occasionally, or what?]

Husband:

While I can do it, I'd like to be able to do it. But there's come a time when you feel as though you won't be able to do it …

Wife:

He won't retire, no … [What about you?]

Wife:

Me? I'll keep on doing what he wants to do. But I want to do other things as well. I'd like craft work and stuff like that. But sometimes I think I'm, you know, sort of, sick of being a farmer. But other times it's OK. It depends. If you're out there and there's fly-blown sheep and the mosquitos are eating you and it's drizzly rain.

Husband:

> The sheep have given us a bit of a headache this year, we've had eight inches of rain for April … [Difficult conditions, yeah. I'm intrigued by what you just said, 'sometimes I feel I'm sick of being a farmer.' That must take a bit of admitting.]

Wife:

> It does. Yeah, I do. I do get sick of being a farmer at times, and I just think, oh well, I'd like to, you know, just, let's go and have a picnic today. But he can't do that because you've got to go round the sheep, you've got to do this, you've got to do something else, and so you don't. Sort of, you get over it the next day.

Husband:

> The weather conditions this particular year have been extreme …

Arthur will not leave the farm until he dies. He blames his father's will for the succession difficulties his son is facing. The thought that he may be part of the problem would not have occurred to him.

End of the Line Farmer

One paddock on Brian's farm was selected by his great-grandfather more than 120 years ago. He has been on it for over 30 years himself, undertaking a range of farming activities, including various forms of livestock and crop farming. Brian has not always been a specialist woolgrower. Other family members used to farm in the district, but Brian is the only member of his family left farming there now. His approach to farming is careful: he doesn't spend money extravagantly. He rotates the stock a bit, but is sceptical about the costs and benefits of cell grazing. Half the farm is leased country and, without the income his wife has brought home from her work in town, things would have been tough.

Brian is not an addicted farmer. He is about 50, and already talks about retirement. Some aspects of woolgrowing he already finds wearying, and he wants time to travel and pursue his hobbies. 'I don't want to be flat out farming when I'm 80 and drop dead, chasing sheep round in circles. I want to do a few things as I get a bit older, well, before we get too old, that is.' Despite this, he is still developing the farm.

Brian's children have all left home to pursue careers off the farm. I asked him who he thought would be running his farm in 20 years' time. 'It won't be me, I'm pretty confident in saying that. And it won't be a member of the family, at this stage. It will be somebody else.' He realises that the family farm, part of which has been in the family since selection, will have to be sold. One extended passage of my interview with Brian is so compelling that it is reproduced here in its entirety, verbatim and unabridged.

[Are any of your children showing signs of going farming?] 'No.' [Is this something you've talked about, or is it just so obvious from their choices that it's not going to happen?]

> Well, we certainly have talked about it, right back from when they were younger. And we made a conscious decision not to try and influence them or persuade them to take on the succession of the family farm, that they could make up their own minds. And, as time's gone on, it just became obvious that farming wasn't an option to them, they weren't considering it as an option. I suppose, during the time that they became more aware of the farming enterprise, it was certainly performing very badly. So it's pretty obvious that they weren't going to choose that as their profession. And, also, we discussed with them the fact that the property wasn't really big enough to support more than one family. So if they were to become interested in farming, it probably wouldn't be on this home property, initially anyway. During that last 10 or 15 year period, the property wasn't big enough to support even one family. It wasn't supporting one family. Virtually since 1990, anyway.

[So, because the farm would still have to support you, any child coming back on the farm wouldn't have enough for themselves as well. That's what you mean, is it?]

> Yes. We've seen it happen, where the child comes back onto the farm. You can see what happened with our friends, as well. And the property's not generating enough for two family incomes. So what tends to happen, is the child will have to get off-farm work, and then they soon become disillusioned with what's going on, because they're trying to run two jobs. And quite often they'll give up after a while and go off and do something else anyway. So, unless something dramatically changes, we'll be the end of the line with the succession on this property. Our children have quite an emotional tie to the property, and they say to us that we must never sell it. But, when it comes to the final crunch, they're never going to be here running it. So eventually it will be sold.

[Yeah, I mean, to tell you that you must never sell it, you know, they're putting some kind of pressure on you, that you know you can't deliver on.]

> Oh well, it's not really putting any pressure on us, because we've made it clear that the family, the property here, if need be, will serve as our superannuation, in order for us to be able to retire. When they say the property should never be sold, it's more to do with them being able to come home on holidays and spend time, you know, bring their friends home, and it's more to do with just being in the country and being in the house here, and it doesn't have much to do with the farming enterprise. We could probably retain the house and a few acres, and still achieve the same result, as far as they're concerned. But it's only their age. As

> they go off and develop their own careers and, you know, and families, and their own lives, it won't be an issue any more.

[How much influence is there in the fact that the property, at least parts of it, have been in the family for such a long period?]

> For me personally, although I like to think that it's not an influence, that we regard the property just as a business, I know deep down that it does have an impact. And, when the day comes we have to sell or decide to sell, it certainly will be quite a wrench. But I'm confident that when the decision is made and we do it, that there'll be no regrets, we'll just move on to the next thing. My mother and father, when they retired, about 15 years ago, they sold the family homestead, where we all grew up, and where my family has been for the last 125 years. And they were sad at the time of the auction, but I think they've hardly thought about it since then. They've got a new house that's warmer and close to the township, and they've just moved on, they're quite happy about the whole thing. So I can see that's just the way it would be for us, too.

Brian is the end of the line.

Second Career Farmer

Claire seems at first like a wide-eyed, naïve dreamer, until you learn how carefully and thoroughly she has pursued her dream. She has felt a call to grow wool all her life and only now, having raised her children, is she doing it. 'I come from a line of farmers going back to five generations ago, coming out from Scotland with their sheep, she says. So I've always had a love of sheep and a love of wool and a love of the country. My father was a farmer.' A family breakdown and upbringing by a mother who hated the country kept her away from farming for a while, but she always tried to stay in contact with the land. 'The wool myth has been in the family forever, and it's always been a major interest.' Does she really reckon it's a myth? 'Oh, it's not a myth, but it's a part of the family legend, I suppose, what everybody sort of talks about, the one thing that everybody has an affinity for. Yeah, it's sort of, it's the thread that goes through the whole, from one generation to the next.' Her love of sheep was obvious throughout the interview. 'They're just delightful animals though, aren't they, sheep, don't you think? They're all different, every one of them's got a different personality.'

Several years ago, Claire and her surgeon husband bought several hundred acres of bare land about an hour's drive on the freeway from Melbourne. The property they chose was, at the time, 'just dust, but very beautiful dust.' They 'fell in love with it and that was that. It was just right.' Closeness to Melbourne was important, so her husband could keep working. Setting up a farm from scratch takes a lot of money, and they wanted to do it properly.

> We wanted to do it, you know, the cutting edge, the very best that we could possibly do it. Find out as much information as we could about, you know, the best fencing, the best way of setting out the paddocks, the best way of water reticulation, the best, you know, all those sort of things. But you need money for that. And the best advice.

They have money from the 'good paddock in town' and several advisers and consultants that they use regularly. To prepare themselves, Claire and her husband did a course for beginning farmers run by Melbourne University and Claire also qualified as a woolclasser. This enabled them to 'ask the right questions.' They employ a manager, and also do a lot of work on the farm themselves.

The dream that Claire is following is her own and she does not expect her children to share it. The children are far away, doing other things. 'This is ours.' And when she can no longer run it, 'then it's gone. Unless somebody in the family wants to prove that they're capable of doing it, or want it badly enough. But no, we don't see it as that. I mean, we've never seen it necessary to organise their lives for them.' She is not likely to make room for them anyway. Would anything make her leave this place? 'Leave it? No, I'm not going to leave, no. Heavens no.'

Claire recognises that she and her husband are regarded by the neighbours as different. It's not just the architect-designed house, or the fancy fox-proof fences. They are newly arrived from Melbourne and have a small farm, in a valley of much larger farms owned by old families. The owners of some of these farms are nearing retirement age, so it is possible that some farms in the valley will be sold soon. The neighbours may not be able to afford to buy them. The farms may well be cut up and bought by more people from Melbourne, few of whom are likely to have a passion for sheep and wool. Meanwhile, Claire tries to fit in.

> We use local people to do things for us all the time. Like cutting hay or pasture renovation, or help at artificial insemination and embryo transfer time, because you need a team of people. So everybody sort of knows what we're doing, and nobody feels that we're a bit secretive or we, you know, these weird people have come from the city and who knows what they're doing all the way down that road. They all talk about us, I'm sure. But it's all positive. They all wander in if they want to borrow something or they want to talk about something. We use the guy next door to transport our wool and bring in grain and all that stuff as well. I think they appreciate that, too.

Claire doesn't have farming forebears or family members nearby to answer to. 'We don't have any baggage as far as farming's concerned.' Apart from a lifelong calling to grow wool.

Part-time Farmer

David's family property was originally settled by his great-great-grandfather. He doesn't have any of the original land himself, but his brother does. David is proud that he chose to go farming, and was not pressured into it. He is also proud that the farm he lives on he bought himself, using money he had saved as a deposit. He knew he would have to buy land of his own. 'The family farm wasn't big enough for myself and my brother and other family members.' His own farm is not large, less than 500 acres, but he wanted to buy within a few minutes-drive of the family property, and prices were high because of demand from Melbourne people even then. With the continual extensions to the freeway, his farm is now little more than an hour's drive from Melbourne. In addition to his own farm, he now owns some of the family property and has recently leased several hundred acres. He knows that he can't afford to buy any more land nearby, says that he was lucky to be able to lease some land for a reasonable price, and is contemplating leasing more if he can find any at the right price.

David had good training for woolgrowing, at agricultural college and as a jackaroo (a kind of farm management trainee). Although he says he likes handling livestock, David enjoys all aspects of farming. His particular passion is doing things properly. He is trying to build up the fertility of his farm with high fertiliser inputs, to enable him to run a higher stocking rate than the district average. David says he is 'still committed to woolgrowing,' but is no longer solely a woolgrower. He has diversified into prime lambs and is considering trading cattle. Few of the Melbourne people, who are the major buyers of land in his area, are interested in running sheep.

David's wife has always worked off-farm, apart from spending a few years rearing the children. Even this income was not enough during the wool price crash of the 1990s, and David himself had to get a job for five years. He was lucky enough to find one that made use of his farm management qualifications. The demands of his off-farm job made it difficult to get the farm work done on time, and this was before he had any leased country.

Several years ago, David put his farm on the market, in the hope that he could obtain enough money for it to be able to buy a much larger property in the pastoral country of the Riverina, about half a day's drive to the north. No one offered enough money to enable him to move his operation, so he and his family are still there. Meanwhile, his four children are growing up and have commitments of their own in the area, and they have extended the house, so the family is becoming more and more settled where it is. Leasing some land has enabled them to expand the scale of the farm.

The one thing that would make David and his family sell up is if one of his children wanted to go farming. Some of them are showing interest, but they are still too young to decide. 'I'd be really happy if any of them wanted to go farming. I'd try and encourage them into it. I don't know how, entirely, yet. But I'd encourage them to do other things first.' And what if all four of them want to go farming?

'Well, they'll all be part-time farmers, won't they? But I think if they're interested, there's always some way of getting there in the end. And it might be, you know, as a part-time occupation.' 'Well', I said to him, 'it has been for you, occasionally.' 'Yeah. Probably will be again one day, if we can't lease any more country.'

Retirement Successor

Ewan was brought up on a woolgrowing property in a renowned fine wool area within sight of the Grampians, in the Western District of Victoria. The farm was only just large enough to support his parents. Ewan's brother was not interested in taking over the farm, but Ewan was. Nevertheless, Ewan had to leave the family farm to earn a living. He went to Melbourne and forged a successful professional career. He married and had a family, but his wife was not interested in farming. The most Ewan could convince her to accept was a few acres on the outskirts of Melbourne.

To satisfy his interest in wool, Ewan got involved in various community activities to do with sheep and wool. He helped to run the wool sections of country agricultural shows, and also judged wool at shows. At the shows, he would talk to me about what was happening on the farm. I was always impressed with Ewan's knowledge of wool and farming in general, considering he had spent most of his life in Melbourne.

Over the last few years, Ewan's father had begun to slow down. He sold off some of the farm and left himself with about 400 acres. This was quite enough for an old man to manage on his own. To further ease the workload, he replaced the fine merinos with cattle. Then Ewan's father died. Ewan and his brother have been taking it in turns to go down to the farm on alternate weekends to make sure things are all right, and to provide emotional and practical support to their mother. His mother would like to move off the farm and into town.

Recently, Ewan and I were talking about his own future. He is now about 60, an age at which many professional people are considering retirement. He is separated from his wife. He has always wanted to be a woolgrower, but circumstances had always prevented it. His father is no longer around to run the family farm, and his mother doesn't want to run it. His brother is not interested in farming. There are no sheep on the property, but he grew up in the district and has kept in touch with the locals. He knows people in the district who would sell him some good sheep. Sure, the farm is not large, but how much land would he need at his age? Surely he has a few good farming years left in him. Perhaps it is time to do something for himself, for a change.

Ewan has some thinking to do.

Discussion and Conclusion

Patterns of entry to sheep farming are changing. Entry by people under 25 years of age is now relatively uncommon, and projections are for the number of sheep farmers under 30 to reduce almost to zero. Young people, even those raised on farms, are choosing to go farming in fewer and fewer numbers, and this is particularly true of the sheep industry. Their reasons relate, no doubt, not only to the low returns that many children of sheep farmers will have observed as they were growing up in the early 1990s, but also to the distasteful work of crutching and dagging sheep and the distance from amenities. Cafes, cultural pursuits, good universities, well-paid jobs with a salary paid into your bank account every fortnight, friends and potential wives are all to be found in the distant city and not the local town. A trend as strong as this will not be reversed merely by a return to higher wool prices. For young people, career and lifestyle opportunities in the city are probably much greater and more enticing than for their parents. This is the life chosen by the children of Brian, the 'end of the line' farmer.

Even if a farm child enjoys the work, staying on the farm may involve too much sacrifice. Children may once have come home on the farm and worked for years for low wages, never knowing when they will be able to take over the farm management, much less the ownership of the farm itself, in the way of the 'farmer's boy' described by Gasson and Errington (1993), but those days are likely gone now for most farm children. The son of Arthur, the 'rusted-on farmer', seems to have more responsibility than a farmer's boy, but is still inhibited to some degree by the constant presence on the farm of his father.

If the farm is not big enough to provide sufficient income for two generations, both the young person returning to the farm and their parents, then the returning young person would have to get off-farm work. The off-farm work available to them would be unlikely to be as remunerative as the work that would be available to them in the city, and any off-farm work they did would detract from the lifestyle benefits of being a woolgrower, particularly the personal autonomy and independence of being one's own boss. Even working on the farm alongside their parents may provide insufficient autonomy and independence. In such a situation, the approach taken by David, the 'part-time farmer', may be preferable: he bought a small farm separate from but close to his family's property and worked part-time when necessary, striking a balance between following his family tradition and being independent from the rest of his family.

Just as many young people are leaving the farm, many middle-aged or older people are returning. Today, mid-life entry, often after a significant non-farm career, has become common. Some entrants, such as Claire, the 'second career' farmer, are following a passion for sheep farming by independent purchase of a sheep property. Others, such as Ewan, the 'retirement farmer' are contemplating rescuing the family farm as their parents age and can no longer cope. It is not easy to tell whether or not, in a given family, this form of inter-generational transfer will occur. Sometimes children who have settled in the city with professional jobs and have said all their adult lives that they will never return to the farm, actually do so when their ageing parents die or announce that the family farm will be sold. Even if this does not actually happen, some parents harbour a wish that it will (Vanclay 2004).

A related phenomenon has been observed in Scotland, in which rural migrants return to the land of their youth in later stages of their working life (Stockdale 2002). It is likely to occur in Australia and will increase in importance. This return of the middle-aged to the farmland of their youth will often be part of the wave of wider amenity migration, rather than a decision based upon the opportunities offered by the family farm. Many of the returning middle-aged will have no need to rely on the farm for much of their income. If they choose to remain in sheep farming, it may not be in a major way. Such returns to rural areas will be more likely in closer settled and high amenity agricultural regions. In these areas, these changes have the potential to create patterns of farm gentrification. In other, less attractive, regions the young will not return in large numbers and population decline will accelerate.

Structural adjustment is a necessary component of any agricultural industry faced with a cost-price squeeze. Although good for the viability of the industry as a whole, adjustment imposes social costs on individuals and families faced with their own adjustment decision. Such costs are greatest for farmers who are forced out of the industry by economic pressures. The form of adjustment that imposes the least social cost and causes the least dislocation is probably the decision made by fewer young people to enter the industry. This is the adjustment that is occurring in the Australian wool industry.

Is the disappearance of the young woolgrower a problem? Does the wool industry need more young farmers? If a young entrant to woolgrowing is to make a reasonable living from it, their farm needs to be large. Because of the cost-price squeeze, the minimum size for a viable farm keeps increasing. There are many small sheep farms in Australia: the financially smallest 40 per cent of them produce only 10 per cent of Australia's total value of sheep production, while the largest 10 per cent produce 40 per cent of the value of production. For a young person growing up on a small woolgrowing farm, even if their parents had been satisfied with the living to be made from it, any expectation that they could build a career based on that farm would be unrealistic. The struggle to make ends meet on an unviable sheep farm is likely to be unrewarding for them and a move to town a better use of their talent and enthusiasm. For the relatively small number of children from large sheep farms there is indeed the potential of a career on the farm. An industry with shrinking labour requirements needs increasingly fewer entrants to maintain itself anyway. For those children from small woolgrowing farms who still want to be involved in the industry, their best option is likely to be to build another career off the farm, then take up farming once their financial security has been assured by their other career. Even if they get into woolgrowing at an older age than did their parents, the increasing number of ageing woolgrowers indicates that they might still be able to have a long career in the wool industry.

References

Anderson, R. 1966. *On the Sheep's Back: Past, Present – Future?* Melbourne, VIC: Sun Books.

Barr, N.F. 2004. *The Micro-dynamics of Occupational and Demographic Change in Australian Agriculture: 1976–2001*, Report No. 2055.0. Canberra, ACT: Australian Bureau of Statistics.

Bell, J.H. and Nalson, J.S. 1974. *Occupational and Residential Mobility of Ex-dairy Farmers on the North Coast of New South Wales. A Study of Alternative Occupations*. Armidale, NSW: Department of Sociology, University of New England.

Blanc, M. and Perrier-Cornet, P. 1993. Farm transfer and farm entry in the European Community. *Sociologia Ruralis*, 33(3/4), 319–335.

Bowsfield, C.C. 1914. *Making the Farm Pay*. Chicago: Forbes and Company.

Buttel, F.H., Jackson-Smith, D.B., Barham, B., Mullarkey, D. and Chen, L. 1999. *Entry into Wisconsin Dairying: Patterns, Processes and Policy Implications*. Madison, WI: Program on Agricultural Technology Studies, University of Wisconsin.

Curtis, A., MacKay, J., Van Norhuys, M., Lockwood, M., Byron, I. and Graham, M. 2000. *Exploring Landholder Willingness and Capacity to Manage Dryland Salinity: The Goulburn Broken Catchment*, Report No. 138. Albury, NSW: Johnstone Centre, Charles Sturt University.

Curtis, K. and Croker, K. 2005. *Wool Desk Report – September 2005*, Wool desk Report No. 6. Western Australia: Department of Agriculture.

Errington, A. and Tranter, R. 1991. *Getting out of Farming? Part Two: The Farmers*, Study No. 27. University of Reading, Farm Management Unit.

Gale, H.F. 2000. Small and large farms both growing in number. *Rural Conditions and Trends*, 10(2), 33–38.

Gale, H.F. 2003. Age-specific patterns of exit and entry in US farming 1978–1987. *Review of Agricultural Economics*, 25(1), 168–186.

Gasson, R. and Errington, A. 1993. *The Farm Family Business*. Wallingford, UK: CAB International.

Gray, J. 1998. Family farms in the Scottish borders: a practical definition by hill sheep farmers. *Journal of Rural Studies*, 14(3), 341–356.

Hawkins, H.S. and Watson, A.S. 1972. *Shelford: A Preliminary Report of a Social and Economic Study of a Victorian Soldier Settlement Area*. Parkville, VIC: Agricultural Extension Research Unit, University of Melbourne.

Hoppe, R.A. 1996. Retired farm operators: who are they? *Rural Development Perspectives*, 11(2), 28–35.

Jackson-Smith, D.B. and Barham, B. 2000. *The Changing Face of Wisconsin Dairy Farms: A Summary of PATS' Research on Structural Change in the 1990s*, PATS Research Report No. 7. Madison, WI: Program on Agricultural Technology Studies, College of Agricultural and Life Sciences, University of Wisconsin – Madison.

Kanel, D. 1961. Age components of decrease in number of farmers, north central states, 1890–1954. *Journal of Farm Economics*, 43(2), 247–263.

Martin, P. 1998. *Profile of Australian Wool Producers*, Research Report No. 98.5. Canberra: Australian Bureau of Agricultural and Resource Economics.

McGuckian, N., Stephens, M., Brown, R. and McGowan, H. 1995. *Your Farm, their Future ~ Together*. Tottenham, NSW: Lachlan Advisory Group Inc.

Napier, R.J. 1972. *Labour use Efficiency in the Sheep Industry*, progress report to the Australian Wool Board. Armidale, NSW: University of New England.

Potter, C. and Lobley, M. 1996. Unbroken threads? Succession and its effects on family farms in Britain. *Sociologia Ruralis*, 36(3), 286–306.

Price, L. and Evans, N. 2006. From 'as good as gold' to 'gold diggers': farming women and the survival of British family farming. *Sociologia Ruralis*, 46(4), 280–298.

Salmon, P.W., Fountain, R.N. and Hawkins, H.S. 1973. *Human Adjustment in Australian Agriculture, 1972: A National Survey*. Parkville, VIC: Agricultural Extension Research Unit, University of Melbourne.

Smith, M.G. 1987. Entry, exit and the age distribution of farm operators, 1974–82. *Journal of Agricultural Economics Research*, 39(4), 2–11.

Stockdale, A. 2002. Out-migration from rural Scotland: the importance of family and social networks. *Sociologia Ruralis*, 42(1), 41–64.

Tolley, G.S. and Hjort, H.W. 1963. Age-mobility and southern farmer skill – looking ahead for area development. *Journal of Farm Economics*, 45(1), 31–46.

Vanclay, F. 2004. Social principles for agricultural extension to assist in the promotion of natural resource management. *Australian Journal of Experimental Agriculture*, 44, 213–222.

Weigel, R.R. and Weigel, D.J. 1987. Identifying stressors and coping strategies in two-generation farm families. *Family Relations*, 36, 379–384.

Wilkinson, R.L. 2009. *Population Dynamics and Succession Strategies of Rural Industry Producers*, PhD thesis. Victoria University [online]. [Available at: http://eprints.vu.edu.au/1943/].

Chapter 4
Intergenerational Farm Business Succession in Japan

Tomohiro Uchiyama and Ian Whitehead

Introduction

The characteristics of agriculture in Japan are perhaps more unusual than most other countries in the developed world. To understand the trends in Japanese farming and related succession processes, therefore, requires a degree of contextual background detail. Three key influences have been responsible for shaping contemporary farming in Japan – industrial development (competition for labour), changing domestic diets (reduction in demand for domestic food supplies), and an increasingly ageing population, the details of which form the rest of this section.

Japan was left devastated in the aftermath of World War II. Its meteoric rise, post-war, is attributed to the competitiveness of its industry in the global market, based on raw materials imports and the export of high-value-added products, the so-called 'added-profit trade'. Although the anchor products have changed, food, fibre, and apparel in the 1950s, steel, shipbuilding, car manufacturing, and petrochemicals from the 1960s, and electronics from the 1970s, the basic structure of the industry has been relatively stable.

Japan's policymakers anticipated this economic growth, and tried to adjust the farming industry accordingly. To cope with the situation, the Japanese Ministry of Agriculture, Forestry and Fisheries (JMAFF) introduced 'the basic law of agriculture' in 1961. The scenario visualized in the law was as follows. With economic growth, many people engaged in agriculture would move to other industries. As a result, the number of farms would decrease. If the remaining farmers could expand their farm size by accumulating farmland from landowners who would leave farming or reduce farm size, it would lead to better productivity and higher income for remaining farmers. Additionally, with economic growth and increasing consumer incomes, a developing market would arise for other agricultural products, in addition to rice, such as fruits and livestock. The result would be an increase in the number of 'viable farms', earning high incomes, as in other industries, and the income disparity between agriculture and other industry would disappear. The restriction on farmland lease or purchase was relaxed and lifted according to the scenario.

Contrary to the scenario, the disparity between farming and non-farming incomes could not be eliminated. There were few 'viable farms', except in some regions and for some enterprises. The disparity between farm and non-farm households was, however, accomplished not by farm business development but by an increase in opportunities for earning non-farm income for those from farm households (JMAFF 1996). One of the policy measures to increase non-farm income in farm households involved facilitating the development of industry in rural areas. As a result, farmers' spouses and children have more easily been able to obtain non-farm jobs without leaving their farms.

Table 4.1 highlights the decline in the influence of agriculture on Japan in the 50 years from 1960. From 1960 to 2005, the area farmed declined by 23 per cent, the number of farm households decreased by 53 per cent and the share of the population in farm households also declined to 6.3 per cent of the population. As mentioned above, increasingly farms have sought off-farm income, with the share of full-time farm households dropping from 34 per cent to 15 per cent. (NB. if a husband farms on a full-time basis and his wife is employed outside the farm, the farm household would be classified as a 'part-time farm' as the household income comes not only from farming but also from non-farm job.) In addition to this, the size of the agricultural labour market has shrunk from 12 million in 1960 to 2.5 million in 2005.

Table 4.1 Trends in agriculture in Japan (1960–2005)

	1960	1970	1980	1990	2000	2005
Total population (million)	94	105	117	124	127	128
Farm household population (million)	34	27	21	17	10	8
Share of population in farm households (%)	36.2	25.7	17.9	13.7	7.9	6.3
Farm households ('000)	6,057	5,342	4,661	3,835	3,120	2,848
Share of full-time farm households (%)	34.3	15.6	13.4	12.3	13.7	15.6
Agricultural labourers ('000)	11,960	8,110	5,060	3,920	2,880	2,520
Farmland ('000 ha)	6,071	5,796	5,461	5,243	4,830	4,692
Agriculture's share of GDP (%)	8.6	4.4	2.5	1.7	1.1	1.0

Source: JMAFF statistics

JMAFF summarised the reasons behind the failure of their policy as follows (JMAFF 1996). First, a large number of agricultural workers moved to other sectors. Whilst the young were more able to sever ties from farming completely, for the older farmers, the instability, and low level of wages from off-farm jobs required them to continue farming, subsidised by off-farm income. Technology progress and mechanization also enabled many part-time farmers to hold on to their farmland and continue rice production. In times of high economic growth, farmland prices also rose, and farmers showed an increasing tendency to realise the asset of the farming property, in many cases capitalising on increasing demand for non-agricultural use. Second, the high economic growth was greater than expected, with growth rates per annum more than 10 per cent in the 1960s. As a result, the income levels in other industries rose remarkably, highlighting the lower incomes from farming, especially rice farming, even after expansion in farm size.

In addition to the structural changes in the labour and property markets, demand for domestic supply was compromised by the liberalisation of imports of agricultural products and consequential impacts on the currency exchange rate, affecting the competitiveness and profitability of domestic agricultural enterprises. Meanwhile, farm performance was also challenged by changes in the dietary habits of the Japanese. With change in wealth, diets of the 1960's based on rice, vegetables, fish, and soy sauce declined (rice by 44 per cent between 1960 and 2000 and soy sauce by 40 per cent), in favour of more non-traditional products including wheat, fruits, meat, milk, and oils (meat + 454 per cent, milk + 324 per cent, and oil + 251 per cent) (Figure 4.1). Ironically, the production of wheat, livestock, and oilseed is not suited to the Japanese climate, with the Asian monsoon climate characterised by hot summer and rainy seasons.

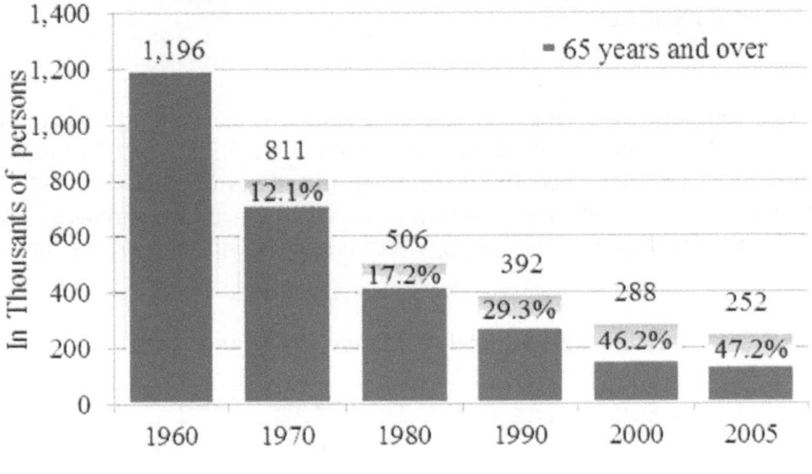

Figure 4.1 Trends in food consumption per person in Japan
Source: JMAFF statistics

To make the situation worse, total food supply in Japan started to decline from the mid-2000s, largely because of an ageing society, an issue in Japan overall, but with specific repercussions in the agricultural sector (Figure 4.2). In 2005, almost half of the agricultural workers were 65 years or older.

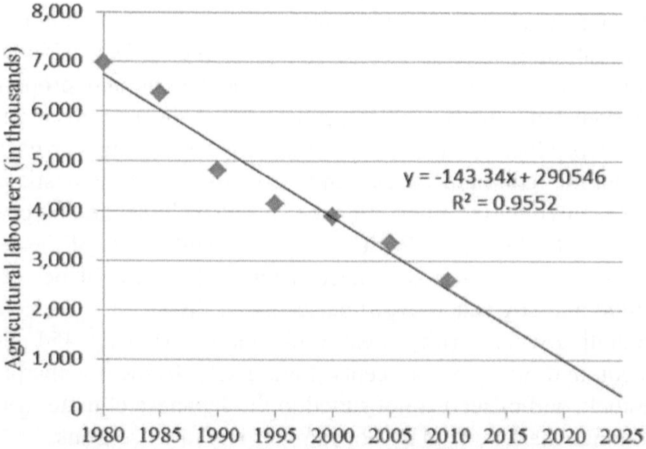

Figure 4.2 Trends in age of agricultural labour in Japan
Source: JMAFF statistics

As a consequence of this, continuing trends leave the share of agriculture in Japan's GDP in 2008 as only 0.9 per cent (JPY 4.4 trillion; USD 44 billion), a decline of 44 per cent since 1990, according to Japan's Cabinet Office (JCAO) statistics. The phenomenon now known as Japan's 'agricultural shrinkage' (some describe it as a 'collapse' [Kudo 2006]) is illustrated clearly in the figures of Table 4.2.

Table 4.2 Japan's 'agricultural shrinkage' (1990 and 2008)

	1990	2008	Difference
Farmland ('000 ha)	5,243	4,628	-11.7%
Farm households ('000s)	2,884	1,750	-39.3%
Price index of agricultural products (2007=100)	118.3	97.7	-17.4%
Gross agricultural output (billion JPY)	11,432	8,474	-25.9%

Source: Japan's Ministry of Agriculture, Forestry and Fisheries (JMAFF)

Trends in the economics of farming over the last 50 years are illustrated in Table 4.3. In terms of consideration of farm succession, perhaps the most significant figure here is the very small mean size of farm units in Japan, increasing steadily since 1960 but, by 2005 still less than 2 ha. Farmland Reform after World War II provides the main reason for such small farm sizes in Japan. Farmers were permitted to own farmland up to one hectare, and were also prohibited from renting any additional farmland. As a result, many tiny holdings were 'born' in Japan. Such size restrictions or lease prohibition no longer exist, but the Reform seems to have had a lasting effect on the ways in which farmers' value farmland – a move from land 'for agricultural production' to land as 'property holding'.

Table 4.3 Farm economics in Japan

	1960	1970	1980	1990	2000	2005
Average farm size (ha)	0.99	1.09	1.18	1.33	1.79	1.85
Income from farming ('000JPY)	219	508	952	1,163	1,085	1,103
Income from non-farming ('000JPY)	192	885	3,563	5,438	4,975	4,323
Farm household income ('000JPY)	411	1,393	4,515	6,601	6,060	5,426
Pension receipt ('000JPY)	32	199	1,079	1,797	2,221	2,286
Gross farm household income ('000JPY)	443	1,592	5,594	8,398	8,281	7,712
Share of income from farming (%)	49.4	31.9	17.0	13.8	13.1	14.3

Source: JMAFF statistics

In terms of Gross farm household income, although this has increased to JPY 7.7 million in 2005, more than 17 times the 1960 level, much of this increase was derived from non-farming income and pension receipts. The average farm household now sustains itself with non-farming income, although this, too, has been in decline in recent years, as the average age of farming householder increases and more and more people retire from non-farming jobs.

The picture developed above is, to some degree, over-generalised, as differences do exist in size, economic contribution from farming, and the time required to run the farm, according to the main enterprises of the business (Table 4.4). Where, for example, rice, vegetable, and fruit production is involved, the average size of holdings is smaller. The average farm economy in general is similar to that of rice farms reflecting the fact that more than half of Japanese farmers mainly produce rice.

Table 4.4 Average farm income per holding in Japan by product (2009)

	Income from farming per farm (1,000 JPY=10USD)	Farm size (ha, head)	Yearly farm working hours by family (hrs)
Rice Paddy	350	1.4	788
Upland Crops[1]	8,600	22.6	3,245
Vegetables	1,770	0.9	2,848
Fruits	1,450	1.0	2,683
Flowers	2,770	0.4	4,837
Dairy	7,680	41	5,437
Beef (breeding)	970	13	2,684
Beef (fattening)	1,600	97	3,290
Pigs	3,960	870	4,331

Note: [1]Hokkaido area
Source: JMAFF statistics

Despite the relatively low income to be derived from rice production, it is worth noting that retention of rice as a crop is partly due to its cultural significance. For example, *Dengaku*, a traditional art in Japan, involving song and dance, is believed to have originated from the twelfth century as a form of prayer for a good crop of rice.

However, an increasing number of farmers are quitting, or decreasing their farm size, and few are willing to rent or buy the farmland released. The land becomes a 'weeds paradise', causing neighbouring farmlands to degenerate as well. Japan's Ministry of Land, Infrastructure, Transport and Tourism (JMLIT) reported in 2007 that 412 rural settlements would disappear within ten years, with another 2,219 possibly disappearing after ten years (JMLIT 2007). This trend is confirmed by JMAFF estimates (2010) that 386,000 ha of farmland have, thus far, been left abandoned.

This section has endeavoured to provide a summary review of farming in Japan, highlighting contemporary pressures and practices, an essential context in which to set the consideration of farm succession in this country. Farming in Japan is unique in many ways, and this has an impact on the importance attached to and the practice of ownership and transfer of business and assets in the industry.

Characteristics of Farm Succession in Japan

Focusing, now, on the issue of new entrants to farming, JMAFF began to maintain statistics on new entrants to farming in 2006 and the results, thus far, are presented in Table 4.5 below. Perhaps not surprisingly, from what has been said before, the overall picture is one of reducing entry, or re-entry into farming, a drop of 17.5 per cent over the four years. 'Back to home farm' means the people who come back to their home farm from non-farming jobs – farmers' children or retired people. 'New employees' are those who are employed by farm businesses for farm work. 'Create new farm' are those who begin their own farm business but do not succeed to any farmland by kinship. From the Table, it is observed that the 'back to home farm' is the largest group, although declining in number in the recent years. With fewer individuals returning to the farm or starting up, it is not surprising, that the number of 'new employees' has risen. New start-ups, or 'Create new farm', are noticeably smaller in number, largely due to constraints on acquiring farmland since they have to obtain permission from both landowners and an agricultural committee in the area, which is usually not easy. Recent legislation requires agricultural committees to make a 'synthetic judgment', with the only criterion on which to make this based on 'harmonization with neighbouring farmers'. There is a degree of anxiety that 'new starters' may not always have the skills to farm at an appropriate level in accordance with this requirement.

Table 4.5 New entrants into farming in Japan

Person	2006	2007	2008	2009
Back to home farms	72,350	64,420	49,640	57,400
New employees in farm businesses	6,510	7,290	8,400	7,570
Create new farms	2,180	1,750	1,960	1,850
Total	81,040	73,460	60,000	66,820

Source: JMAFF statistics

Figure 4.3 clarifies the situation further by showing this same data classified by age. From this it is clear that most 'new entrants' are 60 years old or older, and their main method of entry is 'back to home farm'. This reflects the fact that the recent 'new entrants' are usually not young farmers, but people who retire from non-farming jobs to begin farming by using their home farmland; hence, the Japanese trend to 'retire to farming'. On the other hand, younger people (39 years old or less) tend to be hired by farm businesses or those returning to home farms. The movement of older people into farming can partly be explained by the 'baby boomer' cohort. People born in the late 1940s reach their retirement age (60 years old in Japan) in the late 2000s.

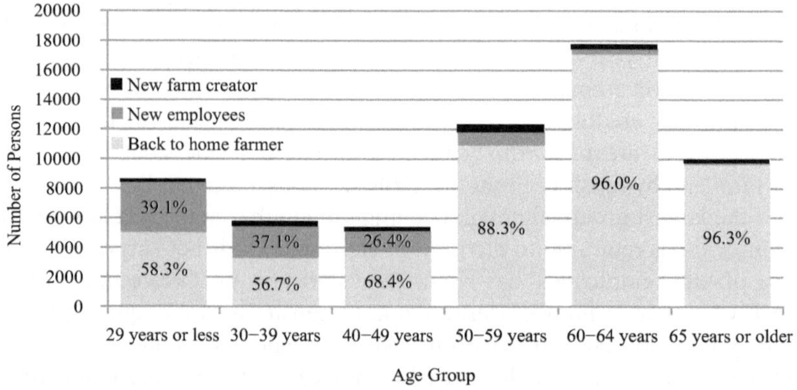

Figure 4.3 New entrants into farming in Japan (2009)
Source: JMAFF statistics

To reveal further unique characteristics of farm succession in Japan, the *FARMTRANSFERS* results are shown in this section. Japan's survey was carried out in 2001 (Table 4.6).

Table 4.6 Farm succession data in Japan: respondents' profiles

	Farm size (ha)	Respondent's age	Successor's age	Number of full-time labours
Mean	4.9	57.1	31.5	2.6
N	4,108	4,136	1,848	4,351
Std. deviation	12.8	10.7	10.5	1.4

Source: *FARMTRANSFERS*

The respondent profiles confirm the foregoing contextual discussion, that is, farms are small, relative to those of many other countries in the developed world, although, in common with the latter, they are run by managers in their late 50's, with assistance from a number of full-time staff.

Identification of a Successor

Clearly, one of the first stages of the succession process involves the identification of a successor. The proportion of respondents in the *FARMTRANSFERS* study who had already identified a successor is higher in Japan than many other countries at 49.8 per cent. For those who indicated that this was the case, the mean age of the successor identified was 31 years (Table 4.6). The likelihood of having identified a successor is in part a function of the age of the farmer, which varies considerably between countries. Figure 4.4 compares the identification of successor by age of respondent and by country. Accepting the trends as the expected norm, it is noticeable that Japan presents a pattern of earlier successor identification than other countries in the survey. This is most probably because, as a result of low farm incomes, Japanese farmers now tend to value farmland more for its potential development value, and less as a means of production. Successors therefore inherit the land with the expectation of realising huge capital returns. The succession is perhaps less dependent on a successor who is interested in farming and the decision is, therefore, more straightforward. In the meantime, as property taxation is lower if the land is farmed, and presumably, for cultural reasons, the land is farmed.

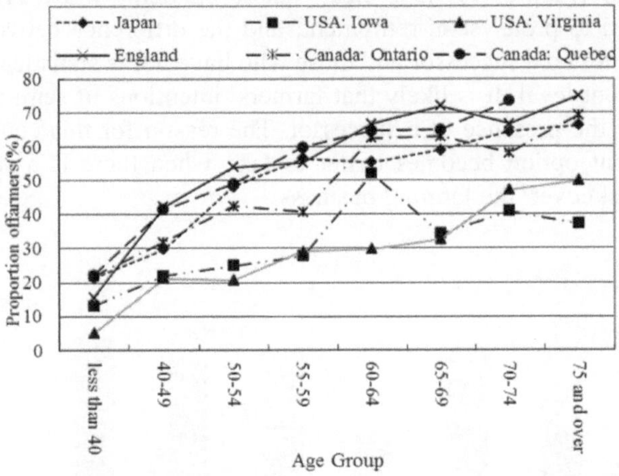

Figure 4.4 Proportion of farmers who have identified a successor by age group

Source: *FARMTRANSFERS*

Farmers' Retirement Plans

Again, perhaps not surprisingly, farmers in Japan appear as the least likely to retire, with 44.5 per cent indicating that they would never retire. Farming life in most, if not all countries, is characterised by the almost inseparable integration of work and home (Table 4.7). For Japan, this is quite clearly extreme, with very few farmers indicating that they will fully retire, eventually.

Table 4.7 Retirement plans of farmers in Japan, USA, Canada, and England

	Japan	USA: Iowa	USA: Virginia	England	Canada: Ontario	Canada: Quebec
Never retire	44.5	27.4	42.1	14.9	22.1	13.3
Semi-retire	37.1	37.6	33.9	47.1	44.0	49.5
Fully retire	18.5	35.0	23.9	38.0	33.8	37.2
N	3,826	391	401	450	529	721

Source: *FARMTRANSFERS*

Analysing the association between these retirement plans and the identification of a successor (Figure 4.5) it is clear that respondents who have identified a successor tend to prefer semi-retirement, and the difference between the group who have identified a successor and those who have not is statistically significant at the 5 per cent level. It is likely that farmers' intentions of semi-retirement are reinforced by the presence of a successor. The reason for this could be that the semi-retirement option becomes easier to take when there is a successor who could 'semi-takeover' the farming business.

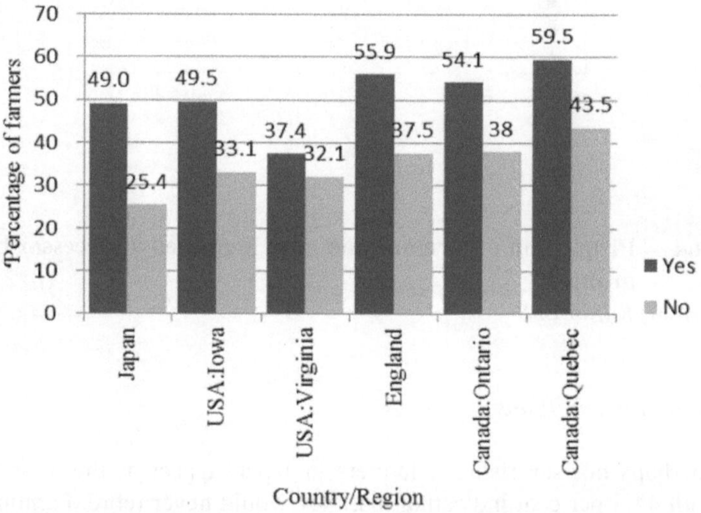

Figure 4.5 Semi-retirement plans by transfer to successor by country/region

Source: *FARMTRANSFERS*

The fact that many people become involved in farming after enforced retirement from off-farm jobs, together with a particular virtue known as *Shogai-Gen-eki* (a tendency for the Japanese, in general, to keep working as long as possible) may explain why such a large proportion of Japanese farmers report that they will 'never retire'. Opportunities in this scenario for 'new blood' entries are, as a result, very few and very, very far between.

Routes to Succession

Once again, the findings from Japan are perhaps unique, this time in terms of the routes to succession. From the contextual discussion in the introduction, it will be of no surprise to see that almost two thirds of farmers entered farming via the 'diversion route', where successors get involved in off-farm employment after leaving school and come back to the home farm at a later stage (Table 4.8).

Table 4.8 Successors' present occupation in Japan, USA, Canada, and England

	Japan	USA: Iowa	USA: Virginia	England	Canada: Ontario	Canada: Quebec
Student	9.4	21.0	13.9	14.2	25.6	26.1
Works full-time on this farm	17.5	24.4	14.8	62.2	33.8	45.2
Works on another farm	1.2	5.0	-	4.1	2.9	1.7
Off-farm employment	60.8	22.7	45.9	4.1	15.9	6.0
Post compulsory education	4.0	1.7	4.1	3.7	6.3	11.4
Manages own farm	0.3	14.3	4.9	6.1	8.2	5.0
Runs non-farm business	5.3	3.4	12.3	4.1	3.4	1.7
Other	1.6	7.6	4.1	1.6	3.9	3.0
Total	100.0	100.0	100.0	100.0	100.0	100.0
N	1,733	119	122	246	207	299

Source: *FARMTRANSFERS*

A number of factors are responsible for this but a key influence is most probably the size of farm business, as smaller farms are likely to provide less opportunity for two generations to work side-by-side. However, it is not entirely clear whether the successors are more involved in off-farm employment because their home farm business size is small or whether the farm business size is small because the successors are involved in off-farm employment. This, alone, cannot explain the much higher percentage of successors with off-farm employment in Japan. What is clear, however, is that in Japan, potential farm successors are able to find off-farm employment reasonably easily and the attractions of this are, no doubt, important. The benefits of this in terms of contacts developed and knowledge and skills, financial and technical, acquired during the 'detour' are potentially advantageous for the returning successor and the farm business.

Transferring Managerial Responsibility between Farmer and Successor

Since effective farm management requires the skills learned and knowledge accumulated from experience related to a particular farm, as well as from formal education and training (that is, a combination of tacit and codified knowledge), the delegation of managerial responsibility is a vital mechanism for transferring farm-specific managerial know-how in a farm business.

Respondents were asked to indicate the extent to which the different decisions had been delegated to the successors working alongside the older generation on the home farm. They did this by assigning a score ranging from one, (the farmers themselves retain full responsibility), to five, (the successors have full responsibility). Scores two to four indicate that the farmer and successor shared responsibility. The analysis presented here classifies eleven items into four decision-making domains (technical, strategic, marketing, and financial) and shows the mean scores of the items that are included in the same group, indicating the degree of managerial delegation in each of these five functional areas. The mean of these five scores indicates the overall level of managerial delegation (Table 4.9).

Table 4.9 Handing over the management 'reins' in Japan

Responsibility Scores		**Technical**	**Strategic**	**Marketing**	**Finance**	**General**
Japan	Mean	2.06	2.09	2.09	1.99	2.03
	Std. deviation	1.06	1.07	1.16	1.01	1.00
	N	909	878	634	765	915
USA: Iowa	Mean	2.12	2.03	1.98	1.87	2.01
	Std. deviation	1.28	1.23	1.30	1.18	1.17
	N	130	127	126	128	131
USA: Virginia	Mean	2.02	1.95	1.87	1.79	1.91
	Std. deviation	1.05	0.94	1.08	0.99	0.91
	N	71	71	71	71	71

England	Mean	2.75	2.50	2.31	1.95	2.38
	Std. deviation	1.18	1.15	1.21	1.00	1.02
	N	181	180	180	180	181
Canada: Ontario	Mean	3.14	3.04	2.87	2.68	2.92
	Std. deviation	1.14	1.16	1.35	1.16	1.09
	N	92	91	89	92	93
Canada: Quebec	Mean	2.94	2.83	2.66	2.40	2.69
	Std. deviation	0.86	0.96	1.11	0.95	0.84
	N	169	161	164	168	170

Source: *FARMTRANSFERS*

The results are largely consistent with the model of the 'succession ladder' by Errington (1998), where delegation begins with the 'technical' and finishes with the 'financial' domain. Table 4.9 reconfirms the existence of the succession ladder (of delegation and responsibility). In Japan, this ladder is noticeably less well elevated than in some other countries, although the difference is not statistically significant. This is most likely related to the small farm size and the high proportion of successors involved in off-farm work. Though this survey has gathered data on some larger farms (by Japanese standards!), this may still explain this finding. A further consideration is the unique situation in Japan, where most farmers are members of an agricultural cooperative, traditionally selling their products through these cooperatives. It might just be that, as a result, the delegation of decision-making in the technical domain is therefore slower than in the marketing domain.

The Unique Characteristics of Farm Succession in Japan

In summary, Japan presents a number of fascinating characteristics in terms of succession in farming. Many of these are clearly defined by the path of development in terms of land tenure, economic development, and the cultural significance of farming. It is not surprising, then, that a very high proportion of farmers never intend to retire and that more successors take the 'diversion' route. The benefits to prospective successors, of what might be described as the 'ongoing professional detour route', are not inconsiderable. Knowledge, skills, and capital obtained off the farm can improve the potential of new entrants considerably. The *FARMTRANSFERS* study has shown that the delegation of managerial authority happens in its own distinctive style, different from that of other countries. Such features present mixed fortunes for Japanese farming. Two of the major issues in terms of succession, not unfamiliar elsewhere in the world, are the steady decline in new entrants coming into farming and the steady increase in age of farmers (Figure 4.6).

Even though the percentage of cases in which successors exist might not be considered low, most farmers, who are in their 70s and 80s, will shortly be expecting these successors to become more actively involved in the holding. In addition, it should be noted that when the respondents answered 'yes' to the question on securing successors, it might mean that the successor would inherit the farmhouse alone, thus separating the farmhouse from the land.

Variations in socio-economic, cultural, and environmental circumstances have a tendency to dictate a range of different approaches by Government and others towards dealing with problems of public significance. Appreciating the decline in number of farming households in Japan in the second half of the twentieth century, and the nation's focus most definitely on industrial development in this period, what efforts have been made by Government and others to respond to these pressures?

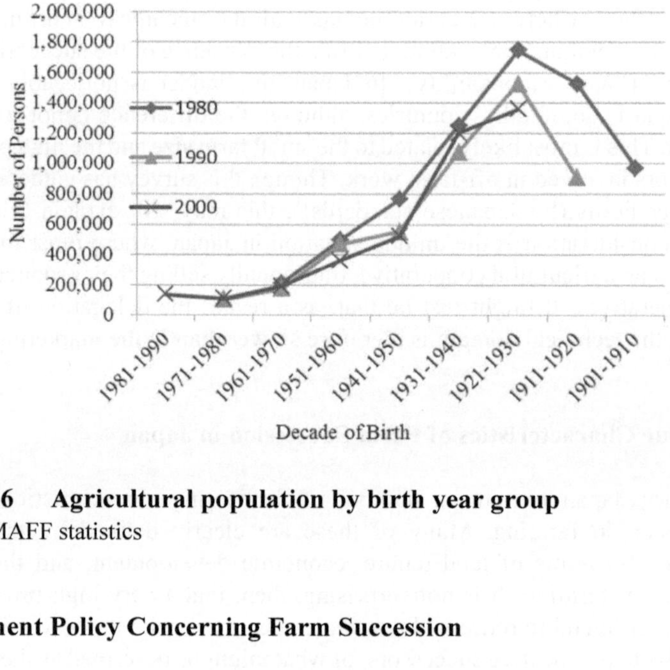

Figure 4.6 Agricultural population by birth year group
Source: JMAFF statistics

Government Policy Concerning Farm Succession

Exemption of Farmland Transfer and Inheritance Tax

Japan's customs and practices on inheritance, including farmland inheritance, are based on primogeniture. However, the current civil law secures the equal rights of inheritance for property in general. If each farm was divided according to the number of heirs, the average farm size in Japan would be substantially less than a hectare. To ensure the viability of farm businesses, therefore, the government provides exemption for farmland from inheritance tax when one heir inherits the farmland solely. According to the rule, if a farmer transfers all the rights on the farmland to his/her successor, the successor is exempted from the transfer tax unless he/she quits farming, sells or leases out, or abandons the acquired farmland. If a successor inherits the whole (or a part) of the farmland and continues farming for 20 years from inheritance, he/she is eligible for full exemption. This rule, therefore, enables successors to avoid the division of the farm holdings, as long as they are willing to continue farming for a long period. The government believes that this rule will help promote farm succession (JMAFF 1996), although, whilst this policy may be useful in preventing further division of farmland, it does nothing to encourage much needed farm amalgamation.

Farmers' Pension Scheme

The Farmers' pension scheme was first established in 1970 to encourage farm succession from one generation to the next. In this scheme, if the pension members retired from farming between the age of 60 and 65, they could get what was referred to as the 'basic' pension. When they retired from farming and transferred their right to the farm to a successor, they would get a further 50 per cent premium. Its objective was as an incentive for older farmers to transfer their farm to the younger generation.

However, the scheme was designed as a pay-as-you-go (self-funding) plan and it collapsed in 2002 as there were more pension beneficiaries than the successors to pay for the pension payments. The successor to this is based on a funded system designed again to correct 'the ageing' problem. All agriculture-related people involved in farming for at least 60 days a year can participate in this scheme. The main feature is that those who meet the requirements benefit from monthly Government contributions, equal to their own contributions, for a maximum of 20 years, as long as they continue to contribute to the scheme. Table 4.10 indicates the detailed requirements for the subsidies. The differential rates in the current scheme clearly encourage young people to take up farming, as they receive higher payments if they meet the requirements before 35 years of age.

Table 4.10 Requirements for government pension contributions under the new scheme

		Subsidies	
		Less than 35 years old	35 years and older
1	An authorized farmer using 'blue return'[1] for tax report	50%	30%
2	An authorized beginning farmer using 'blue return' for tax report	50%	30%

3	The successors or spouse of above the two categories, who have made the family farm partnership agreement	50%	30%
4	Authorized farmer or 'blue return' tax reporter who promises to become the other within three years	30%	20%
5	A successor who promises to become a category one farmer by the age of 35	30%	-

Source: Farmers Pension Fund

Note: [1]The 'blue return' is a tax return where detailed financial accounts are submitted.

Participants become pensioners when they reach 65 years of age and they are entitled to the benefits of the premium when they hand over their farm business to their successors. The eligibility of the successors is the same as in the old scheme. The Farmers' pension scheme was designed to facilitate farm succession by encouraging younger farmers to enter into, as well as older farmers to retire from, farming. Some are sceptical about the sustainability of the new scheme and, in fact, this programme was required to be 'fundamentally reformed' in 2010 by a government revitalization review. The review report concludes that the benefits of the scheme are disappointing, with the number of participants less than 100,000 up to March 2010, against the 2.5 million agricultural labourers in Japan in 2005. It seems probable that fiscal discipline is likely to restrict the level of payments available to participants in the future.

'Farm On Japan' (Farm Succession Aid Programme)

In 2008, JMAFF launched the 'Farm On Japan' programme based on 'the Farm On' programme in Iowa, USA. This programme helps farm businesses by matching 'beginner farmers', who do not own land, with retiring farmers who do not have successors to continue the family farm business (Figure 4.7). In 2010, JMAFF changed the programme name to the 'Farm Succession Aid programme', as it is difficult for Japanese farmers and staff in the related organization to understand the meaning of 'farm on' (!). The programme is funded by JMAFF and operated by the National Chamber of Agriculture (NCA), which has branches in every prefecture (county).

In the programme, both landowners and potential beginner farmers are registered. The beginning farmers can visit the registered landowners to become acquainted with them (trial training of up to two weeks). When the landowner and beginning farmer are ready to enter into an agreement of succession, they enrol for 'takeover' training, which lasts for six months or one year. JMAFF subsidizes the fee for the training. After the training is over, the parties complete the 'succession' according to the agreement.

The NCA highlights several different approaches to succession. In some cases, this involves the sale of the property to the beginning farmer or long-term lease agreements, whilst in others gradual transfers are affected by incorporation.

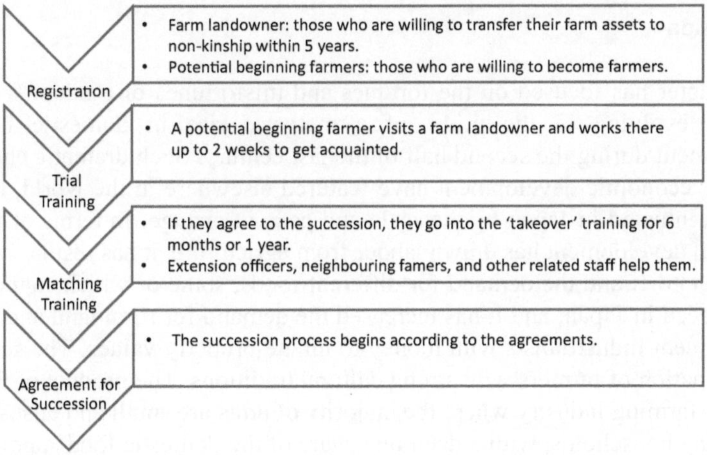

Figure 4.7 Farm succession aid programme
Source: NCA brochure

According to NCA data, in March 2011, of 189 beginning-farmer and 70 landowner registrants, 31 cases were matched for 'takeover' training. Unfortunately, 12 out of 31 cases were cancelled because of disagreement on the conditions, including the value of the property for sale. Under the 'Farm Succession Aid' programme, a coordinator team is formed for each matching case. The members of the team are extension officers, agricultural cooperative staff, and neighbouring farmers. JMAFF and NCA consider it unwise to leave the matching landowner and the beginning farmer to their own devices. A further issue, however, is that, in some cases, the role of 'the coordinator team' has not been clear or there is a lack of leadership in the team. This results in confusion about each one's responsibilities, and the team fails to help the matched cases.

So, the purpose of the programme is to prevent the breaking up and eventual loss of 'excellent' farms with no successors and to promote young people who want to become farmers. Success relies not only on the conduct of the parties and the leadership/facilitation within the programme, referred to above, but also on the availability of landowners who 1) intend to cease farming within five years, 2) are willing to open their balance sheet and profit and loss statement to outside parties, and 3) are willing to transfer their skills and knowledge to 'successors', as well as the farm assets. Traditionally, Japanese farmers have been bound by strict rules to transfer their farmland according to kinship. The legacy of these rules are seen as a major constraint in the success and expansion of the 'Farm Succession Aid' activities, and Japanese farmers may need further assistance to change their ways of thinking in the future.

Conclusion

This chapter has focused on the fortunes and misfortunes of a nation's farming industry evolving in the midst of a meteoric rise in domestic industrial development during the second half of the last century. Such dramatic changes in post-war economic development have featured elsewhere in the world, but have been accentuated in Japan by the scale and pace of change. In terms of impacts, industrial development has drawn labour from agriculture, it has resulted in major changes in diet and the demand for different foods, some of which cannot easily be produced in Japan, and it has increased the demand for rural land and housing from affluent industrialists with money to inflate property values. The setting for this is a nation of farmers with strong cultural traditions. The product of the era is clear – a farming industry where the majority of units are small and unsustainable as farming households, with a declining share of the domestic food supply, partly as a result of the need to liberalise imports; reduced dependence on income from agricultural enterprises on the farm holding (increase in off-farm and on farm non-agricultural income) and a greater tendency for farmers to *retire to farming*, thus perpetuating the ageing issue in Japanese agriculture.

Some fear that Japanese agriculture is facing the threat of 'extinction'. For many years, Japanese agriculture was considered to be stable, although not particularly efficient. This stability was brought about by part-time rice producers who had full-time non-farm jobs. In recent years, however, employment instability and the ageing problem have had a significant impact on agriculture and have left their mark in the rural areas of Japan.

Policy measures have been established by the government in response to this situation. Of them, the 'Farm Succession Aid programme' can facilitate farm business continuity and provide opportunities to potential 'beginning farmers'. This programme provides farmers with a strategy to enter into, as well as to retire from farm businesses without a kinship successor. Although it has the potential for major impact on farm asset utilization and in nurturing beginning farmers, the programme is beset with difficulty in realizing its objectives.

Two potentially long-term constraints to future progress exist in Japan. The first is the structure of farming – the very small holding sizes, insufficient to support a household from farming income alone. This has, perhaps, been perpetuated by the non-agricultural economic development, which has secured a rural economy allowing farming households to subsidise their farming income. The second major constraint concerns the traditional preferences of farmers for holding farmland for cultural reasons or as property with potential development value. The 'territorial hold' of the former is still strong and closely related to tenurial custom in Japan. Whilst there are attractions in terms of preserving links between people and land for a good proportion of the population, there are undoubted issues relating to performance efficiency and the ultimate constraints to a prospective new entrant.

The future holds a number of challenges, not least in terms of food security. Japan's agriculture is likely to pass through a long period of adversity. Both policymakers and academics need to make sustained efforts to assist with the restructuring of agriculture and to continue to encourage retirement and the availability of opportunities for new entrants.

References

Errington, A. 1998. The Inter-generational transfer of managerial control in the farm-family business: a comparative study in England, France and Canada. *Journal of Agricultural Education and Extension*, 5, 123–136.

Japan Ministry of Land, Infrastructure, Transport and Tourism (JMLIT) 2007. Kokudo Keikaku Sakutei no Tameno Shuraku no Jokyo ni Kansuru Genkyo Chosa Saishu Hokoku. (Survey for the Current Situation of Rural Settlements: Final Report for Establishing National Spatial Planning). Available at: http://www.mlit.go.jp/kisha/kisha07/02/020817_.html p.7 (Japanese).

JMAFF 2006. Nogyo Kihonho ni Kansuru Kenkyukai Houkoku (Report of Workshop on the Basic Law of Agriculture). Available at: http://www.maff.go.jp/j/study/nouson_kihon/pdf/report_h080910.pdf (Japanese).

Kudo, T. 2006. *Houkai suru Nihon Nogyo*. (Japan's Collapsing Agriculture), Douseisha, p.181 (Japanese).

Chapter 5
Farm Succession in Switzerland: From Generation to Generation

Ruth Rossier

Introduction

Farm succession is of crucial importance for family farming and the understanding of structural change in agriculture. Both the decision for or against taking over the farm as well as the timing and pattern of entry into or exit from agricultural employment are core variables for understanding agrarian structural change. This chapter reports on a study intended to help clarify the following questions: What is the current succession situation of family farms in Switzerland? What are the most important determinants for a farm either being taken over or given up in the next generation? According to what patterns does the process of succession take place on family farms?

The topography and natural features of the Swiss landscape have exerted a decisive influence on agriculture. The total area of Switzerland is 41,285 km². According to the Swiss Federal Office for Agriculture, agricultural areas and alpine grazing areas together make up more than one third of the total area of Switzerland and, therefore, have a significant influence on the landscape. Grassland accounts for 70 per cent of the utilised agricultural area. Around 25,000 farms grow cereal crops. But the area on which cereals are grown has fallen by 21 per cent since 1996. Twenty-nine per cent of agricultural holdings lie in the mountain region, 27 per cent in the hill region, and 44 per cent in the valley region. Almost two thirds of Swiss farms are devoted to animal husbandry (cows, sheep and goats). The most important products, by value of output, are milk (23 per cent), cattle (11 per cent) and pork (10 per cent). The average farm size is 17.4 hectares. The vast majority of the 60,900 Swiss farms are family-run. These farms, small in international terms, are dependent upon State support. Over the last ten years, Switzerland has in part liberalised agricultural markets, and farmers are compensated for their public services by direct payments. Subsidies account for more than 50 per cent of factor income. This is meant to support sustainable development and take into account the concept of agricultural multi-functionality. The reforms also support the improvement of ecological performance. For instance, in 2008, 9.8 per cent of farms were organic. The largest proportion of organic farms is in the mountain areas, because converting to organic farming is much easier for livestock farms than it is for cropping farms or horticultural enterprises, for example. The reforms have also reinforced structural change, however; between 2007 and 2008 a total of 870 farms closed down (-1.4 per cent). The fall in the number of farms affected both full-time and part-time farms although the proportion of part-time farms remained stable at around 30 per cent. Many of the current farms are too small to guarantee a living for a family. In 2008 average work income per family member working on farms was annually CHF 41,700 and the agricultural income per farm was CHF 64,000. Non-farm income rose from CHF 16,000 per year and farm in 1990 to CHF 24,000 in 2008. The proportion of jobs in the primary sector is 4.2 per cent in 2008. In Switzerland, 170,000 people are working in agriculture (Federal Statistical Office 2010).

The latest age research for Switzerland (for example, Höpflinger and Stuckelberger 1999) reveals a changed picture of people retiring from the labour market. Increased life expectancy has given rise to a generation of older people who are active even after retiring. Once they have left the labour market, these "young seniors" take advantage of their newfound freedom, and develop new lifestyles and leisure behaviour. These new behaviour patterns are associated with a generally improving, good state of health. Generally speaking, the mental and physical well-being as well as the financial and social situation of older people has improved over the past decade. Today, the older generation lives a more tranquil, but nonetheless on the whole more active, retirement than was the case a few years ago. The question is whether the general trends of improvement of older people in society also apply to older people who have worked in agriculture. Conclusions as to the quality of life of retired farmers after farm transfer are arrived at based on the analysis of their activities, their financial situation, their state of health, and their social integration.

Methodological Approach

The survey on which this chapter is based took place in 2004 and was embedded in the international *FARMTRANSFERS* research network. In Switzerland, the written survey on farm succession was intended to elicit the probability rate of farm succession, and to investigate economic and social determinants of farm succession and was used for a quantitative model of Swiss agriculture. As interests and attitudes of the younger generation were not part of the original questionnaire, questions on these were added for the Swiss research study. Similarly, gender dimensions were not included in the original questionnaire. The gendered interest in agriculture and farming succession in Switzerland is thus not part of this chapter but see Rossier, 2008 for further information. As discussed in Chapter 1, the quantitative methodological approach chosen by *FARMTRANSFERS* was designed to fit many different situations and this breadth of coverage has been achieved, to some extent, at the expense of greater depth.

The Swiss study of farm succession is based on the questionnaire of the international research network . Consequently, the questionnaire had to be adapted to the national conditions and the Swiss legal framework. The original English questionnaire was divided for Switzerland into the parts of A and B, and was expanded by the introduction of new sections (parts of C and D). Part A collected general information about the successor and the farm as well as basic questions of farm succession. This part required only minor adaptations and additions to the Swiss context. Part B is only applicable to farms with potential successors. This part had to be adapted more strongly to the Swiss situation, for example some questions made no sense due to the legal prohibition of fragmentation of agricultural property in Switzerland. Some more relevant questions were added, for example whether the farm can offer the successor an existence (economic aspect) and whether a farm succession is still connected to a marriage (traditional aspect). Part C is a completely new section. It was addressed to farm managers without potential successors. The goal was to know why the farm succession was not secured. What does it mean for the present manager? What will happen with the farm where a farmer has retired without a successor? Besides structural data of farms without a potential successor, it is also important to consider the social aspects, in order to develop a better understanding of farm succession. Part D is also completely new. This part is addressed to all 14–34 year old children of the present farm manager. The goal was to find out if, and which of, the other children have an interest in agriculture and in succeeding to the farm and which are not interested. Part D was designed for all children, for designated successors and their non-succeeding siblings. Different statements concerning motivation, interest, and the general image of agriculture had to be evaluated by the young generation.

Switzerland is a multi-lingual county with three official languages: German, French, and Italian. The original English questionnaire was first translated into German, then after pre-test and review, the questionnaire was translated into French and Italian. After translation by different people, the three questionnaires were proof read again for language harmonisation.

The Federal Office for Statistics drew a random sample of 2000 farms from the register of the census of farms (2003) in proportion to the number of farms in each of the 26 cantons in Switzerland, focusing on farm managers from the age of 40 years old (born in 1964 and later).

The response rate after the first dispatch of questionnaires in 2004 was 19.2 per cent (384 questionnaires). After a follow-up and a revision to the sample a total of 776 questionnaires of present Swiss farm managers were available for analysis. This represents approximately 1.2 per cent of all farm managers in Switzerland (1.7 per cent of all farm managers over 40 years). The useable response rate equates to 39 per cent.The additional part D for the younger generation was filled out by 731 children.

The questionnaires were controlled by the Research Station, Agroscope Reckenzholz-Tänikon Research Station and supplied to the Federal Office for Computer Science and Telecommunications (BIT) which undertook the manual data entry.

The sample provides a good illustration of Swiss agricultural conditions with respect to regional distribution, farm size, and the basis on which the farms are run. The average utilised agricultural area (UAA) for the farms was 18.75 ha and an average of 1.33 family members worked full-time on each farm. The average age of the farm managers surveyed was 51.5 years (ranging from 40 to 83 years), whilst their partners were somewhat younger at an average 48.5 years (ranging from 26 to 80 years). The current managers have been running their farms for 21 years on average, having taken over at the age of 30 on average. The farmer's families have an average 2.7 children (zero to eight children).

Part D of the Swiss questionnaire was completed by 213 of the 354 children designated by their parents to be the successors and 509 of the other 2048 offspring recorded in the survey. The response rate of the designated farm successors was 59 per cent, whilst that of the other offspring was 25 per cent. The young men were on average 19, the young women 21 years old.

An additional analysis of the socioeconomic situation after farm transfer is based on fourteen problem-centred interviews with retired farmers from the whole of Switzerland (Agrarian Report 2006). Interviewees were between 62 and 81 years of age. The interviews were held in all parts of the country and in the three official Swiss languages – German, French, and Italian – and care was taken to choose (former) farms of differing production structures. The interviews were recorded, transcribed, and subjected to computer-supported analysis, according to the qualitative content analysis method (Mayring 2002).

Farm Succession in Switzerland

In Switzerland, structural change in farming takes place primarily in the context of intergenerational succession. The farm managers questioned planned to hand over their farms on average between the age of 62 (partial handover) and 65 (full handover). This result is not surprising. In Switzerland, direct payments for farm mangers are paid until the age of 65 when the farm manager will receive the old-age and survivors' insurance (AHV). At that moment at the latest the decision to give over the reins to the next generation or to give up farming has to be taken. A continuation of farming without direct payments is quite unrealistic for typical family farms in Switzerland. In general, most of the farm mangers thus plan to work until retirement age, and don't think of an earlier exit although it happens. Even farm managers without successors are likely to farm until retirement age.

Forty-six per cent of surveyed farm managers stated that farm succession is most likely assured (see Figure 5.1) whilst 27 per cent stated that they did not have a prospective successor. In a further 27 per cent of the cases studied, the question of farm succession was still unresolved. However, the succession situation for family farms becomes clearer with the increasing age of the farm manager. Whereas half of the 40-to-49-year-old farm managers still did not know whether and how things will carry on in the next generation, succession has been settled in most cases once the manager reaches the age of 65.

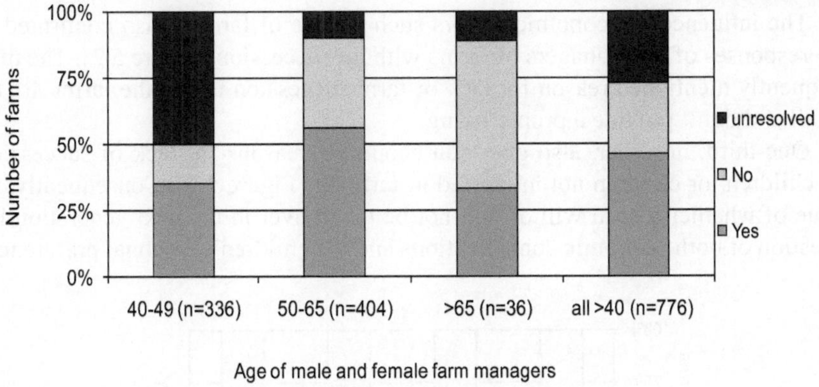

Figure 5.1 Age of farm managers and farm succession

Determinants of Farm Succession

The 213 surveyed farms in which the issue of succession was still unresolved were excluded from the analysis of the causes of succession (leaving a new sample of 563). Three factors were found to influence the likelihood of a family takeover: the size of the farm (area), the number of sons, and the region. The likelihood of takeover increases as farm size (area) and number of sons in the family increases. In addition, farms in the mountain region are more likely to be taken over than those in the valley and hill regions (see Table 5.1).

Table 5.1 Determinants of farm succession

	Valley region		Hill region		Mountain region	
Farm succession	yes	no	yes	no	yes	no
Percentage of farms	60%	40%	61%	39%	69%	31%
Area (hectares)	23.2	14.2	18.1	14.3	20.4	13.5
Number of sons	1.7	1.2	1.9	1.5	1.8	1.4

On the other hand, there is no significant influence exercised on the likelihood of takeover by further farm and family determinants such as type of agriculture, number of family members working on the farm, relative importance of the off-farm income on farms run as a family's main source of income, or age and vocational training of the farm manager.

The influence of economic factors such as size of farm is also confirmed by the responses of the managers of farms without succession (Figure 5.2). The most frequently mentioned reason for lack of farm succession is that the farms are too small and don't provide a proper living.

One-third, however, also give non-economic reasons for lack of succession: no children, or children not interested in farming (Figure 5.3). Consequently, the issue of whether a farm will or will not be taken over in the next generation is a question of both economic considerations and the children's personal preferences.

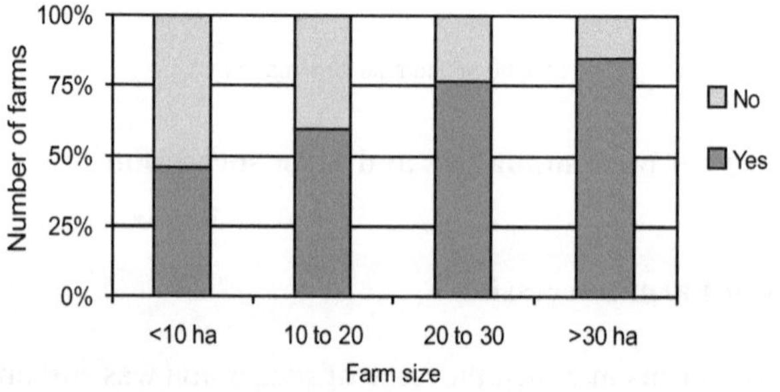

Figure 5.2 Farm size and farm succession

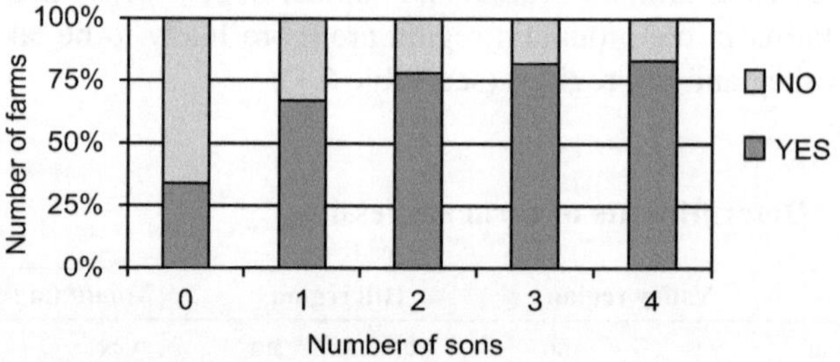

Figure 5.3 Number of sons and farm succession

The Farm-succession Process

Farm succession is not a one-off event, but rather represents a process that goes beyond the formal transfer of goods and chattels to the next generation. The succession process generally takes several years, and follows different patterns (Gasson and Errington 1993). This is also true in the case of Switzerland.

One in three farm managers aged between 40 and 49 years stated in the written survey that the farm succession is settled. However, 48 children were identified by their parents as potential farm successors even though they were under 15 years old. This shows that an initial pinpointing of the farm successor can take place very early on.

Gender still plays an important role in the appointment of the successor. The number of daughters, unlike the number of sons, has no impact on the likelihood of takeover. At 6 per cent, the percentage of potential female successors is as high as that of the female farm managers surveyed (Rossier 2008).

The Pattern of Farm Succession

In Switzerland, two types of succession patterns are typically recognised; to wit, direct entry into agriculture, and entry via professional detour.

Farm successors complete an initial agricultural education, and possibly further agricultural studies/training, leading, for example, to the title of Master Farmer. After completing their training they work full-time in agriculture, as employees on their parents' or someone else's farm, with their own (tenant) farm, or in an intergenerational business partnership with their father.

Before taking over the holding, farm successors complete a non-agricultural education, or are employed outside the agricultural sphere. Thus, they work outside of agriculture for several years until they take over the farm, and often only undergo a second period of agricultural training before the farm is definitively handed over.

Whether direct or indirect entry is chosen often depends on the size of the parental farm. A comparison of the two farm succession patterns reveals that direct entry tends to be chosen by potential successors on large farms, as shown in Table 5.2. Only on these farms is it possible for two generations to earn a living from farming at the same time. The income from a fairly small or medium-sized farm, by contrast, is frequently insufficient even for one generation.

Table 5.2 Occupation of the potential farm successor according to farm size and region

Occupation		All Area (ha)	Valley region Area (ha)	Hill region Area (ha)	Mountain region Area (ha)
Non-agricultural	N=127	16.0	17.8	11.7	12.3
Agricultural	N=140	24.9	26.7	26.4	22.0

Forty-seven per cent of the interviewed managers of farms run as a family's main source of income report having a sideline, either for the farm manager himself, for his partner, or for both. The combination of farming and a skilled part-time job gives potential farm successors two financial mainstays, as well as the ability to react more flexibly to future developments, and the non-agricultural training improves the potential farm successor's chances in the job market. This trend is especially noticeable in the mountain region, where the percentage of farm successors with a non-agricultural vocational qualification, at 24 per cent, is the highest (the figures for the hill region and valley region are 20 per cent and13 per cent respectively).

In most cases of direct entry, farm transfer is preceded by a phase of intensive cooperation, during which the successor is familiarised with work procedures on the farm, and can acquire farm-specific knowledge. In addition to the scope of the cooperation, there is also the question of the extent to which successors can already exert an influence on business decisions. Errington and Lobley (2002) assume that an early and gradual transfer of responsibility results in a smoother takeover. The survey shows that control over financing often remains in the hands of the relinquishing generation until handover. On the other hand, the successors are included in longer-term business plans at an early stage, since fairly large investments must be geared to plans of the successor.

The timing of the handover is nowadays primarily determined by the parent generation's economic situation. The majority of those questioned plan the transfer for when they are 65. Most farm managers are reliant upon their earnings from farming until they reach retirement age. Alternative opportunities for earning a living are usually scarce. At the time of the survey, the average age of the potential farm successors and farm managers was 23 and 53, respectively. If the farm manager hands over his farm aged 65, then the farm successor will be 35 years old at the time of takeover, that is on average five years older than the existing farm managers were when they took over their farm.

Interests and Motives of Potential Successors

Children growing up on farms are tested early on regarding their potential as suitable farm successors. The interest demonstrated by the child in agriculture is taken by the parents as an important incentive to encourage him or her.

For the potential farm successors, activities such as outdoor work and work with animals have the greatest importance. In addition, the practical work and the varied nature of agricultural chores are viewed by over 75 per cent of those questioned as important reasons for taking over the farm. Being able to work independently is another major reason that young people would be keen to take over the parental farm (see Table 5.3).

Table 5.3 Nature of agricultural interest of potential successors and sibling (Two-Sample t-Test)

Statements	Potential succesors Mean Score	Siblings Mean Score	p-Value
I like practical work	1.16	1.89	0
I like working outdoors	1.19	2.04	0
I enjoy working with animals	1.40	2.53	0
I like working independently and being self-employed	1.33	1.85	0
I value the varied nature of farm work	1.25	2.45	0
Continuity of family tradition is important for me	2.15	3.13	
I could see myself working with my parents after a takeover of the farm	1.67	3.27	0
Farming allows me to combine work and family	1.74	2.55	0

Note: Score 1 to 5: agree entirely (1), mostly (2), partly (3), barely (4) or not at all (5)

Also of importance are the family-structure characteristics of the work environment. Well over 50 per cent of all those questioned agree completely that they would like to take over the farmhouse. The compatibility of gainful employment and family and the notion of working with one's parents are also reasons speaking in favour of taking over the farm. Continuation of the family tradition is also cited as a motive, albeit a less compelling one.

The basic economic and political conditions are not rated particularly positively by the upcoming generation. The likelihood of achieving a satisfactory income is viewed especially negatively: just 10 per cent are of the opinion that this is possible. Nearly a quarter of those questioned are fully convinced that they would only be able to achieve a satisfactory income in combination with a sideline. Those interested in taking over the family farm are more optimistic in this respect than their brothers and sisters. The basic agricultural-policy conditions are rated more critically. Job prestige is rated negatively by a majority, with only 25 per cent of the opinion that the farming profession enjoys high social standing. For the farm successors, material interests and the relative social status of the job play a far less important role than their enjoyment of the work.

After Farm Succession

Farm transfer is a process that usually takes several years. It does not always coincide with reaching retirement age. In agriculture, the continuation of work-related activities after official retirement (pensionable age) is the rule rather than the exception. Moreover, retired farmers often continue to work on the farm even after transfer. Should farm succession not take place, they continue to run the farm as long as their health permits. This trend is also borne out by the figures of the Swiss Labour Force Survey (SAKE), an annual survey carried out by the Swiss Federal Statistical Office (BfS): among farmers, 15 per cent of the over-65s are still working. That is twice as many as is the case for independent trades people, for example (see Agrarbericht [Agrarian Report] 2002).

In the agricultural sector, therefore, going into retirement is a gradual process, which gives the handing-over generation time to detach themselves from their life's work. This means that the farming couple do not feel useless in retirement. For the young successor family, the presence of the older generation often means an easing of the work burden, either through help with childminding and housekeeping, or through assistance in barn and field. The passing on of knowledge to the younger generation may also be important (see Agrarian Report 2002). Keeping up certain activities on the farm may also improve the quality of life, and serve to create meaning:

> It's nice to reach retirement age, but it's also a bit too abrupt for someone who's worked all their life, it's also a shock for someone.

> Three mornings a week, I clean out the stables. And I actually enjoy doing it. It makes a change for me. After all, I've got to do something. My wife works from morning until evening, so I can't simply do nothing now.

Although retired farmers' wives tend to withdraw from farm work after succession, they still have their work in house and garden:

> And you, Madam, you've stopped working on the farm?
>
> I have, yes – but he hasn't.
>
> Yes, I've still got my garden. Around the house, no one tells me what I can and can't do any longer. Things are really first-rate now, I must say. We're really content down here now.

Although the workload decreases after farm transfer, taking up new leisure activities in retirement tends to be an exception among the farming population. Retired farmers simply enjoy having more time for everyday things, for themselves or for their partner. Their leisure-activity ambitions are modest, but the activities provide great satisfaction:

> I'm just happy if I can be here in the morning. We [he and his wife] feel at peace together.

Unlike the rest of the retired population, who can take up new leisure activities in retirement (Höpflinger and Stuckelberger 1999), remaining active for retired farming couples means working in their previous field of activity. Working in retirement means genuine quality of life if the work is done voluntarily, that is when activities that are still enjoyed can be indulged in free from duties and constraints.

> I'll tell you – I do what I like, and that's the truth.

> But I've shown [them] how to make gnocchi – I don't mind, because I like to make gnocchi and polenta – [but] I've had enough of the other stuff.

After farm transfer or giving up the farm, the majority of retired farmers continue to live on the farm. This means that they are not entirely free in deciding whether to continue working or not. Out of a feeling of duty on the one hand, and on the other hand because they still strongly identify with their former roles as heads of the farm, the representatives of the handing-over generation feel compelled to continue to pitch in on the farm. They also want to support the younger generation by taking on some of their workload. Tied up with the delight that their own child (or a nephew, grandchild, or neighbour) has taken over the farm and with the idea of carrying on their own life's work, a certain pressure arises which can decisively influence the structuring of retirement, and hence the quality of life:

> We still have the feeling – I do, at least – that I should be helping out here [on the farm].

> And then, he [her husband] has worked long enough as well. Things couldn't carry on like that. And now we have to keep on working (laughs).

Things are different for those who move away from or sell the farm. They have let go, and no longer feel responsible for it. Moving away from the farm opens up new spheres of influence and freedoms:

> If you lease out the farm and he [the tenant] doesn't manage it properly, according to farming legislation you can't simply chase him away. So I thought, I'm better off selling – it will save me grief.

The state pension system was introduced in Switzerland in 1945, and has been effective in combating old-age poverty and safeguarding the economic existence of the elderly. Today, in addition to the state old-age and survivors' insurance scheme (AHV), a majority of the population is also covered by an occupational benefit plan in the form of a pension (2nd pillar) and/or a private plan in the form of an individual pension scheme (3rd pillar). Since 1985, the non-self-employed are actually required by law to subscribe to an occupational benefit plan, though the self-employed are not.

Consequently, as self-employed individuals, farmers are exempt from the obligation to contribute to an occupational benefit plan in the form of a pension. Below, we examine the question of whether retired farmers are missing a second and third pillar, and whether they must therefore remain resident on the farm after its transfer. The time-honoured "right of abode" grants the retired farmers the right to a residence for life and free of charge in specified premises on the farm, provided that nothing to the contrary has been agreed. Whereas in the past the right of abode and remaining on the farm was accepted as a matter of course, nowadays this "right" is contested in both theory and practice (Clemenz 2006).

The state pension scheme (AHV) does not cover the cost of living for retired farmers, especially as only a few draw the maximum pension. AHV contributions are calculated entirely according to income. If farm managers reinvested their profits in the farm (land purchase, machines, farm buildings) and amortised them quickly, this had advantages in tax terms, of course, but the reduced salary also leads to a smaller AHV pension.

> The third pillar? It's in the machinery.

> Quite simply, I need reserves that I can always draw on … The AHV pension wouldn't be enough on its own.

Retired farmers are dependent on an additional income, since liquid funds are in short supply. Entitlement to supplementary benefits is usually out of the question, owing to their financial status (land and real-estate ownership). Some are able to fall back on their savings, whilst others earn a bit on the side, or have rent incomes.

> And then I did a two-year stint nursing for Spitex [nursing and home-care management services]. I gave that up at 70. Now I still do reflexology.

Due to Swiss inheritance laws, the proceeds from the sale of the farm to the younger generation are on the modest side, especially if the farm goes to a child. The Swiss Rural Land Law (BGBB 1991) promotes family farm succession and enables the up-and-coming generation to take over the farm at its earnings value, that is, the capitalised earning capacity of the farm. This can lead to financial difficulties for the handing-over generation, in particular, the older generation of farmers who invested heavily in the farm and/or still have debts to clear, who are left without savings after the sale of the farm.

> [And] when you sold [the farm] to your son?

> It didn't yield anything. Every last franc was used to pay debts. He [the son] still owes me something according to the inventory. He hasn't paid everything yet.

The financial problems become acute for the handing-over generation if the younger generation have no financial resources of their own. The young people are neither in a position to pay the agricultural earnings value of the farm, nor are they considered to be solvent mortgage-holders. It is not unusual for older farming couples to take on a mortgage, so that the younger generation can restructure the farm.

> And then, I said to my wife, I said, we're currently reaching retirement age, I would've liked to stop, and that would be that. To hand over such a large sum at the age of 64, to keep on keeping on even now, I mean, that's what weighs me down a little, and worries me.
>
> Do you have some savings too, then?
>
> Yes, a little. We've got to pay the bank even if we can't pay any more. Because before we paid a lot every year to the bank, and now we don't.

As already mentioned above, the BGBB law facilitates farm transfer for the generation taking over. Since, however, the handing-over generation possesses neither a maximum old-age and survivors' pension, nor substantial savings, those retired farming couples remaining on the farm are still especially dependent on the right of abode. The handing-over generation usually remains in housing on the farm:

> I mean, if we can live here, that's also a part of our income – a large part of it, in fact.

Another option is to move into a rented flat. Reasons for moving away from the farm might include the perceived lack of space for them to remain there, or the sale of the farm. Moving away from the farm and the associated change may have a positive effect on the relationship between the generations:

> I moved because the house isn't built for a large family. The house is already full as it is, with the four children. I wouldn't have had enough space. It's fine this way, though. I wanted to be on my own. We get on very well together.

Studies on the health of retirement-age farmers yield contradictory findings. Steiner (2006) rates the health of the farming population as markedly worse than the rest of the population's. Joint and limb pain owing to years of hard physical labour are particularly common. What's more, emotional serenity is lower, and feelings of loneliness are more common. This is explained by the fact that many retired farmers live off the beaten track. By contrast, based on the data from the Health Survey 2002, Mann (2006) has statistically deduced that there is no significant indication of a precarious state of health among the farming population.

All of the retired farmers who were interviewed ran their own households, and are not yet dependent on outside help. Nevertheless, each interview featured a tale of ill health, with back, hip and, heart complaints being the most frequent. Nor are farmers spared typical symptoms of aging such as high blood pressure, Arthritis, rheumatism, and sudden weak spells. These complaints, however, are accepted as part and parcel of their current stage of life. Some see their health complaints as directly related to the hard physical labour of farming:

> When you look back – we worked hard from a relatively early age – too hard, actually. It takes its toll on you.

> He can't, it must be said, because last year he had to go to the heart clinic, he had an angiogram done, and after that he had a minor stroke which he came out of all right, and the doctor told him not to push himself any longer.

Good health is viewed as a prerequisite for contentment in old age. All respondents felt it was important to maintain a certain independence, as in the continued independent running of their household. Couples count on being able to care for one another, if need be:

> We always hope that one of us will still be able to look after the other, if it works out – that we won't both wind up in need of care at the same time. To stay in our house as long as possible – perhaps with the help of SPITEX [nursing and home-care services].

> I'm now simply content as things are. For me, it's important to be healthy.

The generation interviewed for this research was in some cases still burdened with the "duty of care", in other words, they cared for and nursed their parents or in-laws on the farm until death. This duty is no longer demanded of the next generation. A possibility is an interim solution such as Spitex, which could provide the necessary assistance at home. A subsequent move to a retirement or nursing home is not ruled out, but is postponed for as long as possible.

Before retirement, this generation did not practice recreational sport, on account of both the long hours and physical strenuousness of their work. Now that they are retired, however, many of them feel it is important to exercise regularly. This can take the form of walking, third-age gymnastics, or water fitness classes.

Research into old age makes it clear that it is not just the amount of social contact that is important, but the quality of that contact. Even more important, however, is the emotional attitude (openness, appreciation) towards social networks (Höpflinger and Stuckelberger 1999).

Married couples enjoy their time together in retirement. The many years of working as a team on the farm have bound the couples together. Now, they run everyday errands or pursue leisure activities in each other's company. With widowed or single retired farmers, relationships with children and siblings gain in importance. Married couples also now have greater freedom to visit children who have moved away, or siblings. Being able to cultivate relationships without time pressure is seen as a positive thing.

Whilst some retired respondents increasingly go to a restaurant or café in order to be among other people, others prefer to stay at home. The deliberate seeking out of new social contacts in a club or association, or in the neighbourhood is not noticeable; on the contrary, the farm transfer is taken as grounds for giving up offices or memberships. This withdrawal into the private sphere is viewed as improving the quality of life.

> And when I've got time off, we go out shopping and for a coffee together.

> I was never at home either. I was mayor for 25 years. [Now] I've stopped everything.

Conclusions

The motives for the next generation taking over a family farm have to do with both the farm structures and the non-economic interests and affinities of the successor. In many cases, carrying on the farm is economically feasible only in combination with a non-agricultural sideline.

The size of a family farm has a demonstrable influence on whether a farm is taken over or given up in the next generation. However, since most farm managers are reliant on the income from the farm until their retirement, and many farms are not in a position to support two households, occupational alternatives must be sought, especially in the mountain region. If the intentions voiced in the study are carried out, then we should expect that structural change in farming will continue at the present pace and scale in the medium term. Because of constantly shifting underlying conditions, however, the future of many farms is difficult to predict.

Traditional values and cultural norms such as primogeniture, the appointment of a successor by the father and the moral obligation to take over a farm have lost their importance over the past few decades. This could increase the chances of young women taking over farms in the future. Generally, a fondness for practical activities and the desire to work with animals and to work independently are reasons why young people decide to take over a family farm. Nowadays, the decision to take over a farm is a conscious one, whose advantages and drawbacks are pondered over and weighed up by the future generation.

The findings of the quality of life of the retired farming population show where the opportunities and risks lie in terms of the quality of life after farm transfer, or after giving up the farm.

Retired farmers can certainly be said to lead active lives. Although they arrange their lives along more leisurely lines after farm transfer, they remain involved in the work process if the farm stays in the family. When work is not compulsory and the type of work can be freely chosen, it results in contentment. Work improves quality of life, and gives life meaning. Farmers' wives withdraw from the farm activities, and carry on running the household. Hence, in contrast to the general social trend, an active retirement after leaving farming does not necessarily mean new leisure behaviour.

The financial situation of retired farmers varies, depending on whether the farm is transferred, given up, or leased. The state pension is not enough to cover all living costs. The older generation's right of abode on the farm is therefore an important component of the farm succession agreement between the two generations. Because the farm is passed on within the family at its earnings value rather than at market value, the proceeds from its sale are not sufficient to enable the handing-over generation to create reserves for old age. Consequently, other financial resources must be available in old age. The financial situation of the handing-over generation is generally more secure if the farm is leased to outsiders or sold to someone outside the family. Buildings and agricultural land represent a financial reserve that can be sold if need be, since unlike the remaining population, retired farmers have no second or third pension pillar to fall back on.

The in-depth interviews carried out as part of this study give an indication of how to explain the contradictory quantitative findings on the state of health of the retirement-age farming population. Decades of hard work have left their mark on those who farm for a living. Farmers' health has doubtless suffered under the tough working conditions, and signs of wear and tear can be seen. Mechanisation has made many jobs easier, but even so, a great deal of manual labour is to this day performed in agriculture. The working conditions experienced by the older farmers can be compared with those of builders or HGV drivers: hard physical labour in all weather conditions, and many tractor hours on springless vehicles.

New social contacts are rarely made, but already existing ones are intensified. Retired farmers prefer to enjoy their retirement at home. Having time for themselves, their husband or wife, or their children or siblings is one of the most important features of retirement. In this respect, retired farmers differ from the rest of the population: after a busy working life, it seems to be important to them not to fill their retirement with new activities.

The present analysis shows that the social trend of the changed life situation for the retired population only applies in part for retired farmers. There are risks in old age in terms of financial security as well as state of health. On the other hand, having the opportunity of still being "needed" as a valuable childminder or agricultural adviser, and remaining imbedded in a family network can very much be seen as a chance to live a meaningful retirement.

References

Clemenz, D. 2006. Wohnrecht: Knackpunkt bei der Hofübergabe. *UFA-Revue* 10, 4–6.

Errington, A. and Lobley, M. 2002. *Handing over the Reins: A Comparative Study of Intergenerational Farm Transfers in England, France, Canada and the USA*. Conference Paper of the European Association of Agricultural Economists, Zaragoza, 28–31 August 2002.

Gasson, R. and Errington, A. 1993. *The Farm Family Business*, Wallingford: CAB International.

Höpfinger, F. and Stuckelberger, A. 1999. *Alter Anziani Vieillesse. Hauptergebnisse und Folgerungen aus dem Nationalen Forschungsprogramm*. NFP 32. Bern. 74 pages.

Mann, S. 2006. *Zur Situation pensionierter Bäuerinnen und Bauern*. Z. AGRARForschung, 13(11–12), 506–510.

Mayring, P. 2002. *Einführung in die qualitative Sozialforschung*. Beltz Verlag. Weinheim and Basel. 170 pages.

Rossier, R. and Felber, P. 2007. AGRARForschung, 14(6), 242–247. Federal Office of Agriculture (FOAG), Berne.

Rossier, R. and Wyss, B. 2008. Gender Regimes, Citizen participation and Rural Restructuring. Research in Rural Sociology and Development, Volume 13, 193–216, Elsevier Ltd.

Steiner, A. 2006. Bauernzeitung 47, p. 2.

Swiss Agriculture 2010. Federal Statistical Office FSO, Neuchâtel.

Swiss Federal Office for Agriculture 2006. Agrarian Report. Swiss Federal Office for Agriculture. Bern.

Zürcher Kantonalbank 2006. Analyse des Arbeitsmarktes für ältere Menschen. Chapter 4, pp. 75–120. www.zkb.ch/prospekte/alternde_gesellschaft/pdf/alt_ges_kapitel4.pdf (as at 28.3.07).

Chapter 6

'Keeping the Name on the Land': Patrilineal Succession in Northern Irish Family Farming

Linda Price and Rachel Conn

Introduction

This chapter engages with the issue of patrilineal succession by focussing on Northern Ireland. Across the UK and the developed world succession, it is suggested, has been naturalised as largely occurring from father to son. The word 'patrilineal' is defined by the *Concise Oxford Dictionary* as 'relating to, or based on relationship to the father or descent through the male line'. The culture or way of life of family farming continues to be dominated by the pre-requisite of 'keeping the name on the land' via the male name. Family farms are increasingly struggling to be viable and require the support and work of a growing number of family members, men and women. However, rather than focusing on the 'mechanics' of succession, this chapter sets out to draw on research from Northern Ireland to interrogate the enduring cultural requirement to keep the family on the land via the male line, to fulfil a sense of destiny, belonging and responsibility to past male generations. For men their identity as 'farmer' will be shown to exist in relation to that of women as 'helper'. It is suggested that it is these relational gender identities across generations that enable the farm to survive and patrilineal succession to take place. As the findings will demonstrate, it is this compulsion to 'keep the name on the land' that influences farm decisions, practices and strategies.

The chapter begins by outlining a theoretical basis for patrilineal family farming in the UK which is informed by gender theory. From here, the specific context of family farming in Northern Ireland is discussed before providing an outline of the methodological approach implemented in the research. The discussion then draws on research conducted with existing farm holders and identified successors across Northern Ireland. Here it will be shown that 'keeping the name on the land' remains important both amongst the business holders and those identified as successors. The argument will be developed, therefore, that family farming is patrilineal, that family members work to enable farm survival, thereby facilitating farm succession and that ultimately 'keeping the name on the land' is imperative.

Conceptual Approach to Patrilineal Succession

Agricultural Gender Identities

It is suggested here that farm culture is underpinned by patrilineal, relational gender identities whose construction, enactment and maintenance are becoming ever more difficult to retain. Maintenance of such a culture also requires the compliance of the majority of the extended kinship that comprises the farm family (see Gasson and Errington, 1993) in order for the farm/culture to be perpetuated through the male line which continues to be the norm. Since the 1980s the economic value of women's gendered work roles within the social character of the family farm has been acknowledged (Gasson, 1992; Whatmore, 1991). Political economy perspectives have usefully explained in more detail how the reproductive and productive gender relationships of family farming are integrated in a 'domestic political economy' model. Such work usefully demonstrates the stubbornness of the social structures of family farming. Here farming women are shown to often contribute unpaid work to the farm/culture both inside and outside the farm door in order that the family can avoid being subsumed within capitalist agriculture (see Shortall, 2002; Silvasti, 2003 and Alston, 2006 as Northern Irish, Finnish and Australian examples). Questions arise, however, as to 'how' such gender relations of coping with capitalist agriculture are produced and maintained within the heterosexual 'gender regime' required by the patrilineal culture (Little, 2003). Here Connell's work (2002) is useful in considering the relational nature of farming gender identities through the lens of hegemonic masculinity and emphasised femininity (Brandth, 1995; Morris and Evans, 2001). Thus, dominant norms are shown to exist where the status of farm women complements rather than challenges the status of farming men. Such insights assist conceptualisation of the position of farming men and women around the dominant gender coding of farmer/helper that pervades the culture across and within generations (Price and Evans, 2005; Scott, 1996). From birth, men are usually socialised as a farmer's 'son'and have the opportunity to inherit and learn the skills of farming (Brandth, 2002). Across the developed world daughters have been shown to learn, predominantly, to be supporters of farming men as they move through the life-cycle of 'farmer's son', 'boss farmer' and 'retired farmer' (see Heather et al. 2005; Price, 2010a and Scott, 1996 for Canadian, UK and US insights).

Clearly post-structural approaches to understanding identities of farming individuals resist the primacy given here to the social structures of family farming and the dominant pattern of the farmer/helper dualism. Diverse identities and subjective performance clearly exist in any social setting including that of farming (Butler, 1990; O'Hara, 1998; Bennett, 2005). Clearly, not all sons will want to stay in farming (Ni Laoire, 2005). But it is the overall pattern of farming which this chapter is concerned with and this remains patrilineal. Farm individuals, therefore, are situated within a global agri-economic policy context which requires social relations of production. The experiences of such individuals are also nested within the micro, gendered emotional geographies of the farm. Ideas of hereditary belonging and 'keeping the name on the land', therefore, are clearly important to farm men and supported by the majority of women (Price, 2010a). Despite changes in inheritance laws in countries such as Norway, women's compliance with patrilineality is evident in that it is still predominantly men who inherit, own and retain decision-making on family farms (Almas and Haugen, 1991). Internationally, women have often been shown to suppress their legal and monetary rights, effectively subsidising patrilineal survival (Price, 2006). The status of 'farmer's wife' has been shown to still provide a certain status for many rural women (Hughes, 1997). Research is also beginning to demonstrate the impact the responsibility of maintaining a farming gender identity, the farm itself and ultimately ensuring succession has on men (see Ramirez-Ferrero, 2005 and Price, 2010b as US and UK examples). Here, through their gender identity as farmer it is beginning to be acknowledged how repeated actions of mind and body may lead to men believing a farmer is 'who' they are. For men, as farmer, engage with animals and nature in a spatial arena where legacy, culture, belonging, home and work are intertwined and ultimately reproduced and maintained via the culture of patrilineal succession (Caralan, 2008; Harrison, 2000). Such identities form the bedrock of the culture and thus 'keeping the name on the land'.

Family Farm Survival

As has been demonstrated the farmer/helper dualism underpins the very existence of the family farm. In order for patrilineal succession to occur the farm must survive. However, the farm does not necessarily need to be making a profit. What is important is that the land stays in the family and can be passed down the male line. Even to maintain such survival, given current agri-economic conditions, takes the compliance and efforts of the majority of the wider kinship circle of family farming. Thus, this chapter questions analyses of farm succession that only focuses on family farming as a 'business'. As will be shown in the discussion, family farming is more than this. The land still exerts a hold that dominates family decisions on the trajectory of the farm. It is this understanding that must now, it is suggested, be incorporated into political economy approaches to explaining farm decisions, adjustment and ultimately survival. The political economy perspective became dominant, internationally, from the 1980s onwards to explain agrarian change and its uneven development within the confines of wider economic, state processes (Ilbery, 1998). The approach was subsequently modified to include investigation of the decisions of the 'business holder' in relation to changing, global agricultural reform. Here some acknowledgment of the social relations of the farm household to such development was evident (Munton et al. 1992). The key point here is that approaches to human agency within changing market conditions have largely been confined to a focus on the male 'farmer'. This is mainly as a result of the prevalence of men's appearance in official statistics as business-holder. Such statistics continue to exclude many farming women, sons and retired farmers

(Lobley and Potter, 2004). Therefore, their influence and contributions to farm survival, decisions and succession has lacked acknowledgement (Potter and Tilzey, 2005; Price and Evans, 2006). As has already been shown, without the majority of farming men and women 'signing up' to the patrilineal culture and enacting farmer/helper relational, generational gender identities, it is unlikely that there would even be a farm. This has been underestimated by researchers developing understanding of the motives for adopting strategies which are productivist, post-productivist or multifunctional (Evans et al. 2002; Wilson, 2001, 2008) . For example, decisions on the adoption of farm 'holding strategies', 'expansion' or 'constriction' are likely to be influenced by the availability of a suitable successor and thus options for retirement (Price, 2010a).

Such decisions require the support of the wider farming family. It has been clearly demonstrated that many farming women perform diverse, pluriactive roles in order to support the patrilineal farming culture. Often with the wish to see their sons have the opportunity to take over the farm and carry on imprinting their 'name on the land'. Such multiple roles have been well-documented as including on-farm and off-farm work as well as work which psychologically and financially supports the farm family (Price, 2006; Price and Evans, 2006, 2009). Women have also been shown to take strategic decisions based on the overall motives of the farm culture (Farmar-Bowers, 2010). For example, as Alston (2006) outlines, during the 2006 Australian drought women were prepared to work away from the farm in order to supplement it economically and assist its survival for patrilineal succession. Often such women, as Heather et al. (2005) note in Canada, are well-aware that they are risking their own well-being in carrying such burdens. Farming women are clearly crucial to patrilineal farm survival including, as Gasson and Errington (1993) note, farming daughters accepting pay-offs or 'dowries' on marriage in place of being farm successors. A trend is also noticeable in farming sons often having to wait longer than ever to become farm partners as fear of divorce and farm break up worries the older generations (Price and Evans, 2006). Such complicated family dynamics form the foundations of family farm enterprises, survival and thus succession.

Keeping the name on the land has been shown to be imperative in family farming and this, it is suggested, can only be achieved by the wider kinship circle largely supporting such a way of life. This support coalesces around the farmer/helper identities that across generations farming men and women largely adopt. The context to family farming in Northern Ireland is now provided which is followed by a brief outline of the methodology adopted in the research. The findings will then draw on statistical and narrative data to reinforce how the pre-requisite of 'keeping the name on the land' remains important to existing business holders, influences the choice of successor and motivations for becoming a successor.

Patrilineal Family Farming in Northern Ireland

Agriculture remains important in Northern Ireland (NI) playing a major role in the economy. The Gross Value Added for Northern Ireland in 2009 was 1.1 per cent compared to 0.6 per cent in the rest of the UK (Agriculture in the Home Counties, UK Parliament Briefing Papers 2009). With Northern Ireland being the smallest country in the UK, this shows its continuing dependence on agriculture. Thus, Northern Ireland provides an ideal arena in which to outline the pull of 'keeping the name on the land' via patrilineal succession. This will be shown to be the case despite farm incomes falling over recent years reflecting the trend in the rest of the UK. According to the Department for Agriculture and Rural Development (DARD) in NI, expenditure for farm families for feedstuffs alone has almost doubled in ten years. (DARD, Aggregate Agricultural Account 1981–2010). The income received from animals requiring these feeds has not shown the same returns. Recent figures by DARD, (DARD Farm Numbers, Farm Surveys Branch 2010) show that 75 per cent of farm businesses now focus on cattle and sheep farming, for example (DARD Farm Numbers, Farm Surveys Branch 2010).

According to Agriculture in the Home Countries, (UK Parliament Briefing Papers), 5.7 per cent of Northern Ireland's total labour force is involved in agriculture in general, which is higher than England and Scotland as a whole. The percentage of total land area of Northern Ireland used for agriculture is 73.2 per cent, higher than England which is 68.2 per cent. The average farm size in England is 41 hectares however in Northern Ireland the average is 35 hectares. According to DARD Farm Numbers, Farm Surveys Branch (2010), there are 25,264 Farm Businesses in existence. Of these 89 per cent were classed as small businesses, the majority of which being very small. In 2008 there were 194 fewer farm businesses than in 2007. The downward trend in the number of farms is on average 1.7 per cent per year from 2003 to 2008 and 2 per cent per year over the past ten years (DARD, Statistical Review of Northern Ireland Agriculture 2008). What these figures demonstrate is that family farming, even with falling incomes, remains a key feature of life in rural Northern Ireland. There are currently 29,561 full-time farmers in Northern Ireland, 6,359 less than ten years ago (DARD Farm Numbers, Farm Surveys Branch, 2010). Despite the challenges, the majority of farms in the United Kingdom continue to be run as family farms (see Price and Evans, 2009).

A Farm Business Income above £10,000 was achieved by 67 per cent of the farm businesses in 2009/10; 14 per cent of the farms incurred a loss. (DARD Farm Incomes in Northern Ireland 2009/2010). In contrast with 2008/2009 a Farm Business Income above £10,000 was achieved by 71 per cent of the farm businesses in 2008/09; 11 per cent of the farms incurred a loss. (DARD Farm Incomes in Northern Ireland 2008/2009). Clearly, therefore, it is becoming more of a challenge to remain in profitable family farming.

As Shortall (1999) has demonstrated it is usually sons who take over the farm business. Marriage has been the usual point of entry into farming for women. Traditionally, farms are passed on to the next generation on the owner's death.

The average age of farmers in NI at 58 reflects that across the UK, Cassidy (2004). Anecdotally, it has been thought that a better educated younger generation has been less willing to come into less economically viable family farming that is increasingly hard work as farm workers become unaffordable (see Price, 2010). However, as the findings will demonstrate, keeping the farm going on the land still exerts a pull to stay farming.

The New Entrant Scheme introduced in NI, unlike the Department of Environment, Food and Rural Affairs (Defra) in England, has encouraged Young Farmers under the age of 40 and given financial assistance of an interest rate subsidy on loans to new entrants since 2005. However, the traditions of handing the farm over have remained firmly within the patrilineal culture of sons proving their worth as successors both in terms of skills, but also in terms of having suitable spouses (see Price and Evans, 2006). The increasing inability of the older generation to retire both as a result of loss of farming identity and income is also increasing the time sons spend as 'sons' rather than the 'boss farmer' (Price, 2010b). The New Entrant Scheme (DARD Review of Financial Assistance for Young Farmers Scheme 2010), which commenced in 2004 was only available to farmers under 40 who had been head of the holding (Department of Agriculture and Rural Development Business status) for less than 12 months. The Executive Summary commented on the issue raised with this within the internal review of the Financial Assistance for Young Farmers Scheme by the Department (DARD Review of Financial Assistance for Young Farmers Scheme 2010). However, the review by DARD highlights that the number of farmers under 40 who have acquired head of holding status before this age will be small and decreasing. £4.5m was allocated to Northern Ireland to invest in young farmers. The scheme closed in 2009, however 75 per cent of applicants were approved for the scheme, that is 323 applicants. Fifty-eight per cent of the applicants were deemed as medium to large farms. The number of applicants from farms with Less Favoured Area was low; this may be a reflection on the ability to invest. This explains the low applicant rate from Fermanagh which is almost entirely a Severely Disadvantaged Area. Land in Northern Ireland which is deemed as 'less favoured' by the European Union is given financial aid. Less Favoured according to the European Union is land where 'agricultural production or activity is more difficult because of natural handicap, for example difficult climatic conditions, steep slopes in mountain areas or low soil productivity in less favoured areas.' (Aid to farmers Less Favoured Areas, Rural Development Policy 2007–2013, EU). Findings from the scheme were mainly from areas in the industry where the returns on investment would be most likely, that is larger farms, for example dairy.

In Wales, the Young Entrant Support Scheme (YESS) began in 2009, again for farmers under the age of 40, including a one off grant payment for capital investment for expenses for setting up a holding, advise on training, access to mentors with local knowledge and pairing younger farmers with established farmers (YESS, Welsh Government 2011). In its second year, there has been an increase of 14 per cent of applications and the scheme will continue for five years.

Over £1.5m of grant money has been committed, which is supported by over £2.7m of private sector leverage.

The Scottish Government has modelled their scheme similar to Northern Ireland. The New Entrants and Young Farmers initiative provides interest rate relief on a commercial business development loan and an establishment grant for those who have been in the head of the business for less than 12 months and under 40 years of age (New Entrants and Young Farmers, The Scottish Government). £10m was allocated for the Scottish Scheme alone (DARD Review of Financial Assistance for Young Farmers Scheme 2010).

In England, Defra does not have the same scheme as in the rest of the UK, however it supported Fresh Start in 2004. It aims to help established farmers think about how they can develop their business in the future in the light of CAP reform. (Fresh Start, Food and Farming, DEFRA March 2011). Fresh Start is in contrast to the above mentioned schemes as there is no age limit for the assistance available to new entrants to the agricultural sector in England.

Methodology

The aim of the research on which this chapter is based was to develop a greater understanding of family farm succession planning in Northern Ireland. Specifically this involved considering factors affecting identification of a potential successor and motivations for accepting succession. In order to ensure coverage across Northern Ireland questionnaires to both existing business holders and potential successors were posted to 300 farm businesses. Data regarding farm businesses cannot be retrieved from DARD but those registered can be accessed online using Yell.com. The Yellow Pages is an accepted sampling frame to select respondents for farm surveys and produced 3,169 results for farms across Northern Ireland. The results were sorted into each county, with a random sample of 50 being generated in each county. This was achieved by selecting every fifth farm business from the list provided, 300 in total. The two questionnaires were placed in 300 envelopes, thus 600 questionnaires in total were posted out with two self-addressed envelopes and two covering letters. It is acknowledged that this required the existing business holder to pass the second questionnaire onto an identified successor, where one existed. One hundred and thirteen questionnaires were returned from existing business holders and 75 from successors. The overall response rate was favourable at 31.3 per cent (Mitchell, 1985), 37.7 per cent returned from business holders and 25 per cent from successors.

The questionnaires were designed with a predominance of open-ended questions, with similar issues being put together, in order to derive a broad understanding of the farm businesses and plans for succession across Northern Ireland. The questionnaire for business holders largely focussed on ascertaining what plans were in place for succession, factors affecting selection of successor and conditions and characteristics necessary for succession. The one aimed at the successor focussed largely on attitudes towards being identified as successor and plans for the future.

Anonymity was crucial to the research, with the opportunity being provided to provide contact details if participants wished to be considered to take part in follow-up individual interviews. Providing two self-addressed envelopes ensured that confidentiality was ensured within both groups. Responses were numerically coded and then entered into a spreadsheet. Ordinal data was coded to reflect the order in the data. Similarly, likert scale statements were coded so that five responded to strongly agree and one to strongly disagree statements. The reverse was used for negative statements. Qualitative data was grouped around similar comments. A grounded theoretical approach was taken where themes emerging from the coding and analysis emerged in relation to the overall objectives of the research. These themes informed the topics for discussion in the follow-up semi-structured interviews (Glaser and Straus, 1967).

Five interviewees were selected for follow-up interviews. The overriding themes to emerge from the questionnaires were that the older generation was wary of handing over a 'poisoned chalice' to the next generation who, on the whole, were better educated. However, identified successors, despite falling farm returns, articulated the same attachment and belonging to the land as that of the older generation. This attachment to 'keeping the name on the land' therefore emerged as a key finding of the research. The interviews took place at a location of the participant's choice. Permission was given to record the interviews, but only to aid transcription. Both numerical and narrative data is interspersed in the following discussion. Throughout, the argument is developed that, despite the older generation of business holders having become jaded with farming, staying on the land at whatever age retains an intense pull for farming men.

Findings

Business Holders Perspective (Boss Farmer)

The research was not solely aimed at men. However, it transpired that all those business holders that responded to the questionnaire and thus volunteered for interview were male. Therefore, the findings provide an interesting opportunity to focus on the perspectives of men in family farming from their identity as 'farmer'. This includes being the 'boss' farmer, farmers 'son' or retired 'farmer' as outlined in the conceptual discussion. Of the respondents, 82 per cent had inherited the farm business. As Laband and Lentiz (1983) suggest, farmers do not only have an attachment to the profession but also a desire to maintain the farm in the family as a result of feelings of responsibility to past generations of men. The farms of 32 per cent of respondents had been in the family for four generations or more, with 72 per cent considering it important that the farm remains in the family. This indicates an intense emotional attachment to the land superseding economic imperatives (Jonovic and Messick, 1989). The idea of a succession plan, however, had only been considered by 56 per cent of respondents. This indicates a much more intuitive approach to succession planning based on a feeling of 'when' or 'if' a successor is available and ready to take over. In NI 60 per cent of respondents had identified a potential successor, with 99 per cent being the farmers' son. This confirms the need to keep the name on the land via accepted cultural practices.

However, the route to handing over the reins is becoming more protracted. The majority of business holders felt that acquiring managerial competence took place over a number of years, once trust in the willingness of the successor to keep the name on the land had been assured. Thus, succession is not a single event, but often takes place incrementally over a period of time (see Chapter 1 and also Price and Evans, 2006; Rosenblatt and Anderson, 1981). For 20 per cent of respondents, retirement could not be contemplated until they were 70. Thus, the age of retirement or the inability of farming men to retire completely and thus give up their identity as farmer/male is increasing (Caskie et al. 2002). Giving up identity as a farmer clearly troubled the older generation with age and health being stated as primary reasons for 'having' to retire. This figure is increased where no successor exists. Succession will only be considered by 66 per cent of respondent's when they are no longer able to farm. It is evident, therefore, that allowing succession is an emotional rather than rational process. Such decisions are bound up with a sense of identity, belonging and fear of loss of identity, and attachment if the farm to which that identity is attached is no longer available (Caralan, 2008; Harrison, 2000).

Changes in tax laws and agricultural policy are not shown to impact decisions to facilitate succession. The majority of respondents indicated that such issues are only engaged with 'when' succession happens. The older generation are clearly jaded and tired as a result of ever-greater efforts having been required to keep the name on the land. Falling farm incomes and numbers of farm workers have been shown to increase work burdens and isolation for farming men (Price and Evans, 2009). They worry that whilst they feel they have a responsibility to maintain the farm for the next generation, should they wish to succeed that they may indeed be handing over a poisoned chalice. Worries included: 'Will it be viable to provide a successor with a suitable income?', 'Will it be a millstone around their neck?' and 'Will they have no time off like me?' There is a growing realisation that the farm alone is unlikely to provide an adequate income. However, the majority of respondents indicated how important it is that 'the farm stays in the family', 'we hope that there's enough income to keep a young family, but keeping it in the family that's the main thing'. Existing business holders are aware, therefore, that 'keeping the name on the land' will be increasingly difficult and that it is unlikely that successors will earn a living from the farm alone. Thus, education is not seen as a route out of farming, but rather a way to supplement the farm income. None of the respondents, however, felt that their family should not be able to stay farming, rather that 'there should be funding available for young people' and that 'help should be given to farmers who are reaching retirement'. The 'way of life' of farming is taken as a given, almost a right. As one respondent noted, 'there needs to be better price for milk and beef to encourage younger farmers to be able to pay for extra to give them a better life'. There is a dichotomy here, however, in that business holders feel that government should do more to support the farming lifestyle, but on the other hand that 'there's too much red tape and interference now'. Thus, the autonomy and independence of the lifestyle which has traditionally been valued is felt to be eroded.

What came through strongly in the interviews was that despite the increasing economic pressures on staying on the land that potential successors still value the opportunity to farm and stay on the land. As one business holder stated, 'I tried something else but didn't like it – farming was in my blood'. This sense of being part 'of' the land came across strongly in that 'my farming background would have been the main reasons for my love of animals'. The key message was that farming 'is not just a profession but a way of life'. The naturalization of men as inheritors and custodian of the land featured strongly in that one of the interviewees noted that 'our farm has been in the family for nine generations', but that 'this wouldn't have been the case if there had only been daughters'. Although issues were highlighted where there was more than one son, the findings indicate that these issues are usually resolved early on, so that plans for other siblings can be made. For example, 'it's nearly always the eldest (son) but he wasn't interested in farming, so it went to the next'. Where only daughters exist, as has been noted by Price and Evans (2006), quite often they marry neighbouring farmer's sons. The respondents noted that the likelihood is that 'they'll take the land and incorporate it with the son in laws'.

The overall vision of existing business holders was simply to 'keep going and remain fairly independent'. The younger generation appear to have similar views, discounting some analyses that indicate that younger generations in NI have a more distant relationship with the land (Moss, 1996). Neither do the findings indicate that successors in NI will choose to be part-time farmers (Hennessy and Rehman, 2007). The pull just to 'keep going', to take over the reins and to imprint the family name even more firmly in the land, therefore, appears to be a key concern for potential successors, even if discouraged by the older generation.

Successors Perspectives (Farmer's Son)

Amongst the identified successors, all of whom were male, 57 per cent indicated that they had decided they wanted to be farming before they had left school. Again, attachment to the lifestyle and the land comes through strongly in that 'I always enjoyed farming' and 'I was brought up on a farm and enjoy outdoors and animals'. This again negates the idea of farming purely as profession and confirms that given high land prices that it would be 'practically impossible to start farming without owning a farm'. Enjoyment of farming was the key reason given for wanting to be the farm successor, with 50 per cent having farming qualifications. Clearly the older generation are encouraging greater education, both within and outside of agriculture. As identified successors noted 'there's no money in it now' and 'not enough income to support future plans and family'. Clearly, therefore, just staying in farming and keeping the farm going are accepted as ever-more difficult. A question remains therefore as to 'why' the younger generation would wish to succeed in such unfavourable economic conditions. Thus, the pull of the land and lifestyle clearly supersede economic imperatives.

The theme of 'survival', clearly occupied the mind of successors who just aim to keep the farm going 'as long as possible, try to make a profit although this is becoming more of a challenge'. Clearly successors do want to be farming productively and to 'expand to allow me to survive'. The desire to 'keep the farm in the family' underlined many of the responses of the younger generation and was followed up in the interviews. The majority felt that 'we just need to keep the farming going for the next generation'. It appears that each generation feels this responsibility prior to succession. Again, however, there was no recognition that selling unprofitable farms may be an option reiterating a feeling that 'grants and more funding would help – there should be some sort of financial aid'. Government support is felt to be weak with supermarkets dictating prices when 'better prices for end produce to show the young farmer that the effort and long hours will pay off' were felt to be appropriate. Again, a sense of independence was valued but within the remit of a 'hands off' approach to government support.

Ultimately, the interviews reveal that successors want to inherit the lifestyle and that keeping the farm going, hopefully productively is one way of achieving this. It is 'staying on the land,' however, that is the greatest pull. As one identified successor noted: 'It wasn't expected or even wanted, my parents wanted me to become a doctor or a vet'. It appears, therefore, that each generation of potential successors feels a pull to maintain the lifestyle of family farming and to 'keep the name on the land'. It seems that this potential to continue the patrilineal lineage exerts a strong hold. As one interviewee indicated 'it's a great environment, great for children – town children know nothing about the like of it – it's a great start for them – it brings a bit of closeness between the children and their parent'. Having the opportunity to provide children with the experience of growing up on a farm appears important. Many of the respondents indicated that 'it's all I wanted to do – it's in your blood'. Interestingly, the older generation appear to be encouraging the younger generation away from farming, but feel a responsibility to keep it going until the opportunity to succeed is provided. One interviewee stated that 'I felt no pressure to follow my father', another that 'I was encouraged from childhood to pursue an alternate career'. Successors appreciate that 'it isn't an easy life'. It is this circle of responsibility that came through amongst both generations, to provide the 'opportunity' if the next generation should want to take it up.

The fact that the majority of successors had no major plans for the farm is interesting; merely 'we just want to keep the farm going'. This indicates the importance of lifestyle over profit. There is awareness that the farm alone will not provide an income and that part-time work and subsidization by farming women will often be required (Price, 2006). The feeling that 'they', the urban society, do not understand 'us', the rural/agricultural one, was prevalent in the responses. Successors want to stay farming, want government support, but also want the freedom to enjoy their lifestyle. As one potential successor indicated, 'farmers should be left to get on with the job they do best, feeding the country'. This desire to ensure a 'rural' upbringing for children appears to be part of the pull of 'staying on the land' against the increasing encroachment of urban society (Scott et al. 2007). Such a rural/agricultural culture is felt to be in jeopardy, but giving the next generation the experience to grown up on a family farm is clearly important. It is not difficult to see how this responsibility to just 'keep going' for the next generation to circulate such experiences takes hold. So, successors indicate that they will 'just keep going' and 'keep our heads above water'. Despite awareness across the generations of the increasing hardships of the lifestyle and understanding of the closeness to nature, family and history is implicitly understood. Thus, keeping 'name on the land', whilst no longer easy or even encouraged, retains a pull across existing business holders and their potential successors. The findings confirm that it is this emotional pull to keep going, to protect the lifestyle and to ensure its existence for the next generation to experience that enables patrilineal family farming to continue in its present form.

Conclusions

Two key conclusions can be drawn from the findings. Firstly, existing business holders feel a responsibility to provide the opportunity to keep the farm in existence in case their successor, usually a son, wishes to succeed them. There is clearly a conflict emerging in the emotional landscape of existing successors. When men reach their mid-fifties they often appear jaded with the lifestyle. The increased work involved with falling profits and loss of farm workers has ground down their enthusiasm. Often they are tired of the work and of having to rely on farming women, in particular, just to keep the farm going to enable succession. Education of farming sons is being encouraged but whilst the older generation often hope that their children will adopt professions outside of farming, they still feel an intense responsibility to keep the farm going in case their sons 'wish' to succeed. They do, however, wish the farm to be kept in the family and not sold, even if it does not provide an income. Increasingly, this is fulfilling their sense of responsibility to their forefathers. So, on the one hand the older generation are fed up with the lifestyle, but on the other feel that they had the opportunity and so must their sons. Keeping the name on the land is important, therefore, but the older generation would be content if the major family income did not come from agriculture. Thus, there appears to be a love/hate relationship with farming life. On one side the patrilineal opportunity is appreciated, for it is this that provides farming men with their dominant identity. On the other side there is a sense that it 'traps' older men into the identity of 'farmer'. The prospect, therefore, of leaving or retiring from farming when so much of self, mind and body feels an 'attachment', a sense of 'belonging' to the cultural and physical space of family farming appears to be a huge challenge for older farming men.

Secondly, it can be concluded that despite encouragement to move away from farming by their fathers, identified successors still appear keen to carry on the lifestyle, even when the farm cannot provide an adequate income. The respondents confirmed the same sense of belonging to the land and the family story that their fathers felt. However, whilst their fathers have often become jaded with the life it is clear that their sons still have the enthusiasm to be part of the family's 'rootedness' in the soil and of 'keeping the name on the land'. The sense of being men through whom the patrilineal line travels has, almost inadvertently, been sewn from birth. There appears to be something about growing up on the farm that leads men to often imbue a sense of pride of being born to farm, a sense of destiny, of it being in their blood. This is clearly hard to pull away from. Potential successor's feelings of pride and place in the world came through in the findings. The findings show how an affinity to the landscape, the land and nature are all important to farming men with the 'sights' and 'sounds' mingling almost unconsciously with everyday geographies and actions. This is what Raymond Williams might describe as a 'structure of feeling' (Williams, 1972). All of the male respondents felt that farming is 'in my blood'. Such intense feelings are powerful, producing lived realities. Providing the opportunity for the next generation to experience the farming lifestyle was imperative to the identified successors. These less tangible aspects of farm life have been largely ignored in research considering farm adjustment and survival strategies. It is suggested here, however, that greater focus on the historical, lifestyle aspects of family farming should be acknowledged in explaining family farm survival. As the findings demonstrate, profit is *not* the key motivation for wanting to take over the family farm. In a rural location, being the 'boss farmer' still provides men with a certain status. To be largely autonomous, surrounded by nature and 'fixed' by a patrilineal destiny still exerts a powerful hold for the men taking part in the Northern Irish study. Here being a 'rural/farming' individual, working with the land and nature is still shown to be thought of as preferable to an urban lifestyle.

The findings are illuminating, therefore, in reinforcing the 'pull of the land'. Of course the respondents had all stayed in farming and research with those who have chosen to leave or have no inheritors is required. Farm survival and succession has been shown to be increasingly difficult to achieve and requires increasing efforts by the majority of family members. Farming women have been shown to be crucial to this enterprise, but to largely support the 'idea' of patrilineal succession. It is suggested that future work on strategies for farm survival would be enhanced by considering how 'keeping the name on the land' influences survival and management strategies of family farms. As many of the respondents said 'we just want to keep going!' It is this 'keeping going' via patrilineal succession for motivations deeper than profit that now requires greater attention.

References

Agriculture in the Home Counties, UK Parliament Briefing Papers 2009 [accessed June 2011], www.parliament.uk/briefing-papers/SN03994.pdf

Aid to Farmers in Less Favoured Areas, Rural Development Policy 2007–2013, Agriculture and Rural Development. European Commission EU [accessed July 2011], http://ec.europa.eu/agriculture/rurdev/lfa/index_en.ht

Almas, R. and Haugen, M. 1991. Norwegian gender roles in transition: the masculinization hypothesis in the past and in the future. *Journal of Rural Studies*, 7, 1, 79–83.

Alston, M. 2006. The gendered impact of drought. In B. Bock and S. Shortall (eds) *Rural Gender Relations: Issues and Case Studies*. CABA, Wallingford, pp 165–181.

Bennett, K. 2005. The identification of farmer's wives: research challenges in the northern fells, Cumbria. In J. Little and C. Morris (eds) *Critical Studies in Rural Gender Issues*. Ashgate, Aldershot, pp 60–75.

Brandth, B. 1995. Rural masculinity in transition: gender images in tractor advertisements. *Journal of Rural Studies*, 11, 123–133.

Brandth, B. 2002. Gender identity in European family farming: a literature review. *Sociologia Ruralis*, 42, 2, pp 181–201.

Butler, J. 1990. *Gender Trouble: Feminism and the Subversion of Identity*. Routledge, London.

Caralan, M. 2008. More-than-representational knowledge/s of the countryside: how we think as bodies. *Sociologia Ruralis*, 48, 4, pp 408–423.

Caskie, P., Davies, J. Campbell, D. and Wallace, M. 2002. *An Economic Study of Farmer Early Retirement and New Entrant Schemes for Northern Ireland*. Queen's University, Belfast.

Cassidy, M. 2004. *Tapping into the Rural Heart*, BBC News website [accessed July 2011], http://news.bbc.co.uk/1/hi/northern_ireland/3499511.stm

Connell, R. 2002. *Gender*. Polity Press, Cambridge.

Department of Agriculture and Rural Development, Policy and Economics Division, *Statistical Review of Northern Ireland Agriculture* (2008) [accessed April 2010], http://www.dardni.gov.uk/stats-review-2008-final.pdf

Department of Agriculture and Rural Development Aggregate Agricultural Account 1981–2010, Policy and Economics Division (2011) [accessed June 2011], http://www.dardni.gov.uk/index/publications/pubs-dard-statistics/pubs-dard-statistics-agricultural-aggregate-account.htm

Department of Agriculture and Rural Development Farm Numbers Survey Farm Census Branch 2010 [accessed April 2010], http://www.dardni.gov.uk/index/publications/pubs-dard-statistics/pubs-dard-statistics-farm-numbers.htm

Department of Agriculture and Rural Development, Review of Financial Assistance for Young Farmers (2010) [accessed April 2010], http://www.dardni.gov.uk/full-report.pdf

Department of Agriculture and Rural Development Farm Incomes in Northern Ireland 2008–2009, Farm Census Branch (2010) [accessed June 2011], http://www.dardni.gov.uk/index/publications/pubs-dardstatistics/farm_icomes_in_northern_ireland__2008-2009.htm

Department of Agriculture and Rural Development Farm Incomes in Northern Ireland 2009–2010, Farm Census Branch (2010) [accessed June 2011], http://www.dardni.gov.uk/index/publications/pubs-dard-statistics/farm-incomes-in-northern-ireland-2009-10.htm

Department of Agriculture and Rural Development; Agricultural Census Historical Labour data 1912–2010 (2011) [accessed June 2011], http://www.dardni.gov.uk/index/publications/pubs-dard-statistics/agricultural-census-historical-labour-data-1912-to-date.htm

Evans, N. 2009. Adjustment strategies revisited. Agriculture change in the Welsh marches. *Journal of Rural Studies*, 25, 2, pp 217–23.

Evans, N., Morris, C. and Winter, M. 2002. Conceptualizing agriculture: a critique of post-productivism as the new orthodoxy. *Progress in Human Geography*, 26, pp 313–332.

Farmar-Bowers, Q. 2010. Understanding the strategic decisions women make in farming families. *Journal of Rural Studies*, 26, 2, pp 141–151.

Fresh Start, Food and Farming, Department for Environment Food and Rural Affairs DEFRA 2011. [accessed June 2011], http://www.Defra.gov.uk/food-farm/farm-manage/training-and-new-entrants/

Gasson, R. 1992. Farmer's wives – their contribution to the farm business. *Journal of Agricultural Economics*, 43, 1, pp 74–87.

Gasson, R. and Errington, A. 1993. *The farm family business*. CAB International, Wallingford.

Glaser, B. and Strauss, A. 1967. *The Discovery of Grounded Theory: Strategies for Qualitative Research*. Aldine de Gruyter, Chicago.

Harrison, P. 2000. Making sense: embodiment and the sensibilities of the everyday. *Environment and Planning D: Society and Space*, 18, pp 497–517.

Hennessy, T. and Rehman, T. 2007. An investigation into factors affecting the occupational choices of nominated farm heirs in Ireland. *Journal of Agricultural Economics*, 58, 61–75.

Hughes, A. 1997. Rurality and 'cultures of womanhood': domestic identities and moral order in village life. In P. Cloke and J. Little (eds) *Contested Countryside Cultures: Otherness, Marginalisation and Rurality*. Routledge, London, pp 123–138.

Ilbery, B. 1998. *The Geography of Rural Change*. Longman Limited, Harlow.

Jonovic, D. and Messick, W. 1989. Psychological issues in planning for the farm owner. *Journal of Financial Planning*, July, pp 137–141.

Laband, D. and Lentz, B. 1983. Occupational inheritance in agriculture. *American Journal of Agriculture Economics*, 65, 2, pp 311–314.

Little, J. 2003 'Riding the rural love train': heterosexuality and the rural community. *Sociologia Ruralis*, 43, 4, pp 401–417.

Lobley, M. and Potter, C. 2004. Agricultural change and restructuring: recent evidence from a survey of agricultural households in England. *Journal of Rural Studies*, 20, 4, pp 499–510.

Mitchell, T. 1985. An evaluation of the validity of correlational research conducted in organizations. *Academy of Management Review*, 10, pp 192–205.

Morris, C. and Evans, N. 2001. Cheese makers are always women: gendered representation of farm life in the agricultural press. *Gender, Place and Culture*, 8, pp 375–390.

Moss, J. 1996. Pluriactivity and survival? A study of family farms in Northern Ireland. In S. Caruthers and F. Miller (eds) *Crisis on the Family Farm: Ethics or Economics?* Centre for Agricultural Strategy, Reading.

Munton, R., Marsden, T. and Ward, N. 1992. Uneven agrarian development and the social relations of farm households. In I. Bowler, C. Bryant and M. Nellis (eds) *Contemporary Rural Systems in Transition: Vol 1: Agriculture and Environment*. CAB International, Oxon, pp 61–74.

New Entrants and Young Farmers, The Scottish Government [accessed June 2011], http://www.scotland.gov.uk/Topics/farmingrural/SRDP/RuralPriorities/Packages/NewEntrantsandYoungFarmers

Ni Laoire, C. 2005. 'You're not a man at all!' Masculinity, responsibility, and staying on the land in contemporary Ireland. *Irish Journal of Sociology*, 14, 2, 94–114.

O'Hara, P. 1998. *Partners in production? Women, farm and family in Ireland*. Berghahn Books, Oxford.

Potter, C. and Tilzey, C. 2005. Agricultural policy discourses in the European post-Fordist transition: neoliberalism, neomercantisism and multifunctionality. *Progress in Human Geography*, 29, 5, pp 581–600.

Price, L. 2006. A new farming subsidy? Women, work and family farm survival. *Royal Agricultural Society England Journal*, Stoneleigh Royal Agricultural Society, Warwickshire, pp 49–58.

Price, L. 2010a. 'Doing it with men': feminist research practice and patriarchal inheritance practices in Welsh family farming. *Gender, Place and Culture*, 17, 2, pp 81–99.

Price, L. 2010b. The damaging impacts of patriarchy on UK male family farmers. In D. Winchell, D. Ramsey and R. Koster (eds) *Sustainable Rural Community Change: Geographical Perspectives from North America, the British Isles, and Australia*. Eastern Washington University Press, Washington, pp 42–62.

Price, L. and Evans, N. 2005. Work and worry: farm women's way of life. In J. Little and C. Morris (eds) *Critical Studies in Rural Gender Issues*. Ashgate, Aldershot, pp 45–59.

Price, L. and Evans, N. 2006. From 'as good as gold' to 'gold diggers': farming women and the survival of British family farming. *Sociologia Ruralis*, 46, 4, pp 280–299.

Price, L. and Evans, N. 2009 From stress to distress: conceptualising the British family farming patriarchal way of life. *Journal of Rural Studies*, 25, 1, pp 1–11.

Ramirez-Ferrero, E. 2005. *Troubled Fields: Men, Emotions and the Crisis in American Farming*. Columbia University Press, New York.

Rosenblatt, P. and Anderson, R. 1981. Interaction in farm families: tension and stress. In T. Coward and W. Smith (eds) *The Family in Rural Society*. Westview Press, Colarado.

Scott, A., Gilbert, A. and Gelen, A. 2007. *The Urban Rural Divide, Myth or Reality*. Macaulay Institute, Scotland.

Scott, S. 1996. Drudges, helpers and team players: oral historical accounts of farm work in Appalachian Kentucky. *Rural Sociology*, 61, 2, pp 209–22.

Shortall, S. 1999. *Women and Farming: Property and Power*. Macmillan Press, Basingstoke.

Shortall, S. 2002. Gendered agricultural and rural restructuring: a case study of Northern Ireland. *Sociologia Ruralis*, 42, 2, pp 160–175.

Silvasti, T. 2003. Bending borders of gendered labour division on farms: the case of Finland. *Sociologia Ruralis*, 43, 2, pp 154–167.

UK Agriculture, http://www.ukagriculture.com/uk_farming.cfm

Whatmore, S. 1991. *Farming Women: Gender, Work and the Family Enterprise*. Macmillan, London.

Williams, R. 1972. *The Country and the City*. Chatto and Windus, London.

Wilson, G. 2001. From productivism to post-productivism … and back again? Exploring the (un)changed natural and mental landscapes of European agriculture. *Transactions of the Institute of British Geographers*, 26, pp 77–102.

Wilson, G. 2008. From 'weak' to 'strong multifunctionality: conceptualizing farm-level multi-functional transitional pathways. *Journal of Rural Studies*, 24, 3, pp 367–383.

YESS Welsh Government Farm and Scheme Information 2011 [accessed June 2011], http://wales.gov.uk/topics/environmentcountryside/farmingandcountryside/farming/youngentrantsupportschemeyess/?lang=en

Chapter 7
Non-successional Entry into UK Farming: An Examination of Two Government-supported Schemes

Brian Ilbery, Julie Ingram, James Kirwan, Damian Maye and Nick Prince

Introduction

Most chapters in this book examine different aspects of retirement, inheritance and succession in family farming. As they show, intergenerational transfers are a fundamental element of family farming systems and the transfer of business ownership and control to the next generation is often crucial to the continued development of the business (see also Brookfield and Parsons 2007). This chapter is different to these other contributions in the sense that it is concerned with non-successional entry into UK farming and examines two attempts to attract 'new blood' into an industry that is increasingly characterised by an ageing farm population and, by inference, a lack of innovation and competitiveness. As reported by Defra (2006), 68 per cent of the UK farming population was over 55 in 2005 compared to 49 per cent in 1990, with the median age of farm holders increasing from 55 to 58 over the same time period. Part of the explanation of this ageing process relates to low rates of entry into farming by new, young and non-family farmers (ADAS 2004). The main barriers to enter farming for non-successional entrants are the poor availability of land and high start-up costs (Williams 2006), inextricably linked to low rates of exits, especially from within the family farm sector. Indeed, the introduction of the Single Payment Scheme (SPS) in 2005, as part of the reforms of the Common Agricultural Policy (CAP), has arguably made it possible for some (who might otherwise retire) to remain on the farm without actually farming it. Thus, interventions that enable 'new blood' to enter the industry are important in a farm policy environment that increasingly recognises the need to increase agricultural productivity and develop sustainable systems of production to meet the concomitant challenges of climate change, food security and land resource management (Lobley and Winter 2009).

The difficulties associated with non-successional entry into UK farming cannot be examined in isolation from broader processes of change within the agricultural industry. Thus, market forces, changing patterns of land ownership and structural change all impact on the composition of the farming population. In relation to structural change, e.g. UK agriculture has experienced successive rounds of restructuring in which the main trend has been towards fewer, larger and more capital-intensive farms. However, as Lobley and Potter (2004) demonstrated, the process of structural change is geographically differentiated and influenced by a range of factors such as farm type, farm tenure, location and developments in the wider economy. Crucially, structural change has led to a reduction in the number of holdings available, and a low turnover of farms to let, on the open market, both to the distinct disadvantage of potential non-successional entrants (Ilbery et al. 2010a). Lobley and Potter (2004) found that only 8 per cent of their farm sample in six study regions in England consisted of new entrants to farming. This figure is higher than an entry rate of just 2 per cent in UK agriculture found by ADAS (2004) between 2000 and 2004. Both figures are considerably lower than a suggested exit rate of 18 per cent as the total number of farms and farm businesses have declined. Such figures led Drummond et al. (2000) to talk about a structural crisis in British agriculture, although Lobley and Potter (2004) intimated that the process of disengagement from mainstream agriculture is more subtle, prolonged and spatially differentiated than a crisis perspective would suggest. The current policy emphasis on food security, and its inherent productivism, reflects this sense of spatial and temporal differentiation, in that family farms, in some places at least, may once again be encouraged to concentrate on 'mainstream' agricultural activities (if indeed they ever left them).

Nevertheless, processes of structural change are related to changing patterns of land ownership and a continued squeeze on available land for non-successional new entrants into farming. While the family farm has proved quite resilient to the growth in private land ownership, both at a relatively large (e.g. professional people including farming in their portfolio of business interests) and small (e.g. lifestyle hobby farmers) scale, this has had a negative impact on the supply of agricultural land for rent by new entrants. Indeed, stagnation in the provision of tenanted land, caused mainly by a system of Full Agricultural Tenancies (FAT) in which tenant farmers had a long-term agreement with their landlords and, often, rights of succession, led to the introduction of Farm Business Tenancies (FBT) in England and Wales in 1995 (Whitehead and Millard 2000, Whitehead et al. 2002, Ilbery and Maye 2008, Ilbery et al. 2010b). While the intention of FBT, with their shorter-term agreements and fewer succession rights, was to release more land for letting and so improve the chances of non-successional entrants gaining land, the prospects for new entrants have not really improved. This is because FBT agreements often cover relatively small amounts of land, over short periods of time, and involve 'bare' land without a house and/or farm buildings. Thus, they have been used by established farmers, including family farms, to expand their businesses; this has further restricted opportunities for new entrants. Not surprisingly, Whitehead et al. (2002) showed that less than 10 per cent of FBT went to new, non-successional entrants.

Clearly, if left to natural forces, the opportunity for non-successional entry into UK farming will continue to be reduced. Yet, significantly, most agricultural policies have focused on early retirement schemes, especially in mainland Europe, rather than on new entrants to farming. Thus, there is a need to examine related policy interventions that support non-successional entry. The main objective of this chapter, therefore, is to examine two schemes in the UK that have attempted to provide opportunities for new farmers to enter agriculture: the long-established County Farms Estate (CFE) and the more contemporary Fresh Start (FS) initiative in Cornwall in the south-west of England. It is recognised that, while seeking similar goals, both schemes have operated in very different ways, over different periods of time, and in the context of contrasting policy backgrounds. Thus, the CFE has operated at the national scale while FS has had a more local (county) focus; likewise, the CFE has used its assets to 'set up' new farmers, whereas FS has effectively been working without any real assets. In both cases, and in contrast to the intergenerational transfer of assets in family farming, there is no transfer of assets as such in the CFE and FS.

A case study approach is adopted in this chapter. However, given the varying nature of the two schemes, the examination of the CFE and FS will also be different. While the section on county farms adopts an essentially historical perspective based mainly on the analysis of reports and secondary data, but with some primary data input, the evaluation of FS is based on primary data gathered from interviews and focus groups with key stakeholders, delivery partners, industry representatives, consultants and FS applicants in Cornwall. The intention is not to provide directly comparable analyses of the two schemes, given relative differences in terms of scale and operational mandate, but to draw out more general messages from both scheme evaluations in the context of debates about the family farm. The next two sections provide short cameos and evaluations of the two schemes, before the final section assesses some of the implications of the findings in relation to non-successional entry into farming and provides a general conclusion.

The County Farms Estate

Initially introduced under the Small Holdings Act of 1892, the fundamental aim of the CFE was to provide an opportunity for new, non-successional farmers, and especially young people, to enter agriculture in an affordable manner and to develop viable agricultural enterprises (ACES 2007). Under the original act, a small holding was defined as 'land acquired by a Council … which exceeds one acre and does not exceed 50 acres' (Small Holdings Act 1892: Part I). Crucially, this act imposed a statutory requirement on individual County and Unitary Authorities (CUA) to provide smallholdings (county farms) for farmers where a need existed; it also contained a compulsory purchase clause to ensure that land became available for such provision. After a slow start, the CFE had extended to 80,600 ha and 14,908 holdings in England and Wales by 1914. Growth continued after the First World War, accounting for 177,265 ha, or just over 1.4 per cent of agricultural land in England and Wales, and 29,532 holdings by 1926. Perhaps worryingly, over 60 per cent of these county farms consisted of just bare land.[1]

Both Smith (1946) and the Wise Committee Report (1966) highlighted the significance of the CFE especially in the eastern region of England, notably in Norfolk, Suffolk, Cambridgeshire, Bedfordshire and south Lincolnshire, as well as in parts of the south-west (Somerset, Wiltshire and Gloucestershire), south Wales and Cheshire. Key factors in this development included the availability of land and labour, good soils for the development of intensive farming, and good transport links, especially railways, to the major conurbations of London and Birmingham. Of course, the growth of the CFE coincided with the sale of up to 30 per cent of land in England and Wales between the 1870s and the 1930s (Cannadine 1990). These sales also coincided with the lowest financial returns for land for over a century (Newby et al. 1978). More recently, Winter (1996) argued that, despite expansion, the success of the CFE was limited because wider structural changes in land ownership in the UK did not benefit smaller-scale farms and agricultural labourers.

Further growth in the CFE was limited by the state of the economy and the Second World War. After the war, smallholdings were seen as part of an integrated agricultural policy under the Agricultural Act of 1947. As a consequence, the CFE underwent a process of farm amalgamation and thus witnessed a continual reduction in the number of county farms. Initially, as more county farms became full-time businesses, the fall in numbers (just over 20 per cent by 1964) was much greater than the relative size of the total CFE. However, under the Agricultural Act of 1970 it was no longer the 'duty' of CUA to provide smallholdings, although they 'shall make it their general aim to provide opportunities for persons to be farmers on their own account' (Agriculture Act 1970: Part III). The act also stipulated that the CFE should be reorganised by amalgamating and enlarging some of the existing holdings, as well as creating new county farms. This quickened the pace of restructuring as both the number of county farms and the extent of the CFE started to fall quite rapidly. The intention of creating more economically-viable holdings was sound, especially when set against a wider national trend of fewer, larger holdings and increased land ownership (Northfield Report 1979). However, a number of CUA used the restructuring process as an opportunity to dispose of their county farms, taking advantage of a rising and very different land market from the one that yielded poor financial returns in the 1930s.

Table 7.1 summarises the decline of the CFE in England and Wales between 1964 and 2007. It can be seen that the fall in number of county farms, by over 70 per cent, has been much more dramatic than the decline in the total area of the CFE, by about one-third. Accordingly, the average size of county farms more than doubled between 1964 and 2007; however, at just 26.2 ha this is much smaller than the average size of farm in England and Wales, at around 57 ha. Geographically, Cambridgeshire has the largest CFE in England, with around 14,000 ha and just

1 Bare land is defined as land let without the provision of either a farmhouse and/or outbuildings.

over 250 farms, together with Suffolk (Eastern England), Cheshire[2] (North West England) and parts of the West Midlands and the South West (e.g. Gloucestershire). However, in relation to the area of tenanted land more generally, the CFE is poorly represented in northern regions, the South East and parts of the Midlands such as Nottinghamshire and Northamptonshire.

Table 7.1 Changes to the CFE in England and Wales, 1964 to 2007

Year	Area of CFE (ha)	No. of holdings	Average holding size (ha)
1964	177,883	16,346	10.9
1974/5	164,725	9,823	16.8
1996/7	133,922	5,316	25.2
2001/2	126,188	4,684	26.9
2006/7	117,705	4,488	26.2
% change	-33.8	-72.5	+140.4

Sources: Wise Committee Report (1996); Northfield Report (1979); CIPFA statistics (2002–2007)

Focusing on the more contemporary period of 1997 to 2007, significant losses in the area of the CFE occurred in three clusters (Figure 7.1): first, in the south east of England (e.g. Essex, Kent and East Sussex); secondly, in north eastern areas of England (e.g. North Yorkshire, Durham and Northumberland); and thirdly, in north-west England (Shropshire and Lancashire). These losses seemed to involve either large-scale disposals (as in Kent, East Sussex and, more recently, North Yorkshire) or gradual and regular losses that were consistent with the sale of property as and when freehold possession became available (e.g. Essex and Northumberland). The greatest losses of land within an individual CFE occurred in Shropshire, where regular disposals saw the estate decline from 2,366 ha in 1997 to just 797 ha in 2007. In contrast, some growth occurred in the area of the CFE in parts of southern England (e.g. Hertfordshire, Hampshire and Surrey), either side of the Humber estuary (North Lincolnshire and the East Riding of Yorkshire), and in north and south Wales (e.g. Gwynedd, Flintshire and the Vale of Glamorgan) (Figure 7.1). However, these relatively small gains were easily outweighed by the overall losses in the area of the CFE.

2 Cheshire County Council ceased to exist as of April 1st 2010, becoming two independent local authorities: Cheshire East Council, and Cheshire West and Chester Council. The county farms estate service is currently maintained under the banner of Cheshire Farms Service.

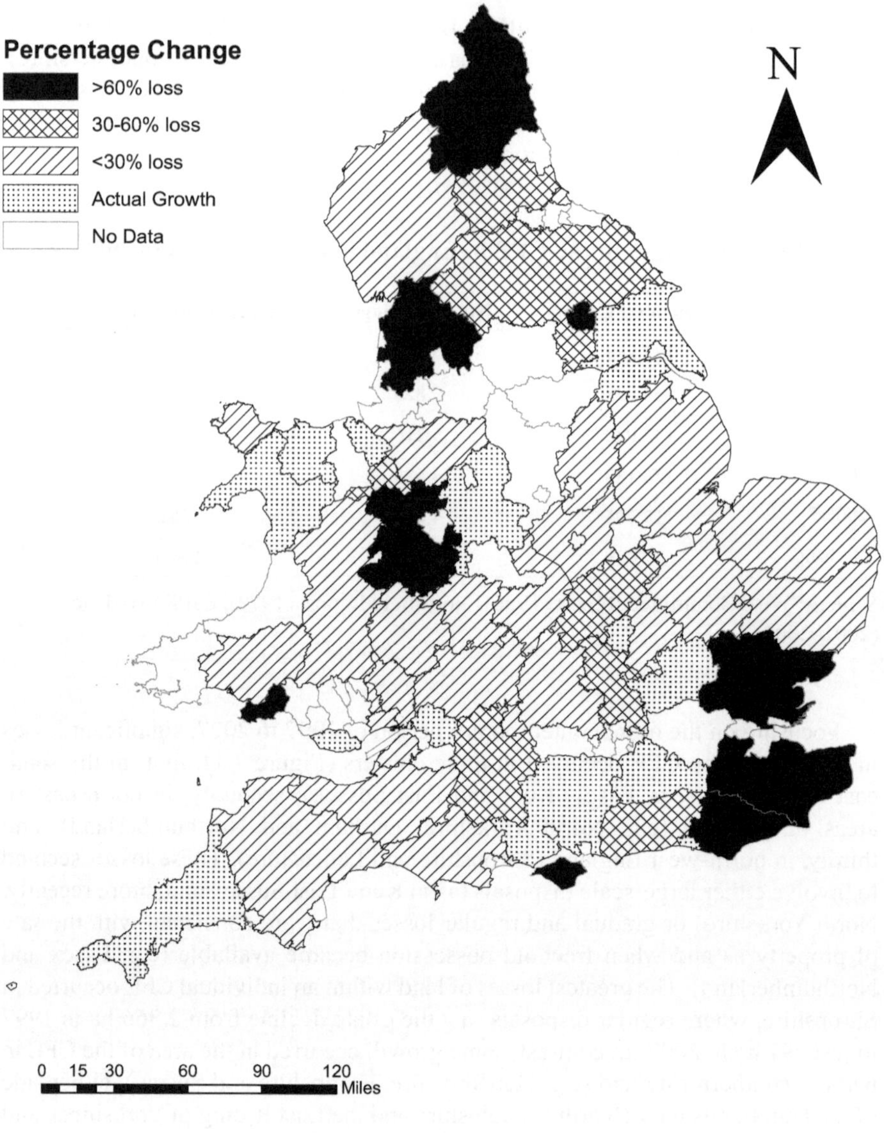

Figure 7.1 Percentage change in the area of the CFE in England and Wales, 1997 to 2007

Two other key features characterised the CFE between 1997 and 2007. First, there was a considerable reduction in dairy farming, in comparison to some growth in arable and mixed farming. Reasons advocated for this decline include the economic realities and difficulties facing the UK dairy industry more generally, the implementation of Nitrate Vulnerable Zones (NVZ) policy which placed additional economic burdens on both tenants and landlords, and the high capital costs of renewing and/or extending fixed equipment for milk production. Local authorities have also found it just as easy to let non-dairy holdings that provide equal and sometimes higher rental returns, thus reducing the economic incentive to provide dairy holdings. Secondly, statistics released by the Chartered Institute of Public Finance and Accountancy (CIPFA 2007) showed a fall in the provision of equipped holdings and a growth in the letting of bare land holdings. The latter is again regionally specific and tends to be a feature of arable farming in eastern England where, e.g. over 40 per cent of CFE holdings in Lincolnshire and Norfolk in 2007 were bare land holdings. This contrasts with more livestock farming areas, where in Cornwall and Powys, e.g. only one and two of its 109 and 170 holdings respectively comprised just bare land .

Thus, today the CFE accounts for only around 1 per cent of the agricultural area in England and Wales and about 3 per cent of the tenanted area in England (just over 5 per cent in Wales). Nevertheless, it retains some importance for a number of reasons. For example, the Central Association of Agricultural Valuers (CAAV 2007) showed that in 2006 county farms accounted for just over one-third of all new let, fully-equipped holdings; research by the current authors in 2008 also demonstrated that over 40 per cent of all new tenancy agreements on the CFE by responding local CUA were granted to new tenants. This provides evidence that the CFE still presents an important opportunity for 'new blood' to enter farming. Earlier research by Errington, Millard and Whitehead (1998) had shown that 24 per cent of those starting up non-inherited agricultural businesses were utilising county farms, just as Whitehead et al. (2002) demonstrated that one-third of new county farm lets were to new starters and that 90 per cent of these were for fully equipped holdings. In addition, rent returns from county farms in England and Wales realised over £20 million in 2006/7, with an operating surplus of around £10.8 million, and in the same year the CFE employed nearly 6,000 people. However, and despite such statistics, county farms continue to struggle in a number of respects, especially in relation to wider trends in land ownership and land tenure in the UK. Increasing owner occupation and the renting of additional bare land by larger-scale farmers mean that first time entrants to the farming industry will nearly always be outbid for tenanted land (Ilbery et al. 2010b). The limited number of equipped holdings for rental (both county farms and private rental) is another potential obstacle as existing landlords are looking to use the farmhouse and/or outbuildings for non-agricultural activities.

Further investigations reveal that the problems of the CFE run deeper than this. A report by CIPFA (2007) suggested that in 2006 only ten county farm tenants took up farming opportunities elsewhere and thus released the land for either new entrants or the enlargement of existing county farms. Thus, stagnation within the CFE is a real problem, a situation made worse by the relatively high average age (55) of county farm tenants, with a number in their 80s and over. This lack of progression, an originally important function of the CFE, reflects a lack of desire by tenants to move and/or a lack of equipped holdings to rent within the wider agricultural sector. It also reflects an attempt by some county farm tenants to negotiate longer-term FBT than is normal in the tenanted sector as a whole. Indeed, some local authorities value the continuity that retaining tenants bring to their estate, even when this discourages progression. Cambridgeshire, e.g. has been keen to retain good tenants who understand localised farming requirements; as a consequence, they have offered them longer term tenancy agreements. CAAV (2007) suggested that the average length of FBT offered by the CFE has increased since 2001, to between five and 10 years; this compares with an average duration of a FBT of 3.75 years in the wider agricultural tenanted sector. Nevertheless, the number of lifetime and retirement tenancies in the CFE has been falling, especially among those CUA making increasing use of short-term agreements on land with an identified development potential. Indeed, some CUA expressed an interest in selling off the CFE in order to benefit from rising land prices (NFU 2007). Thus, North Yorkshire adopted a blanket policy of selling off county farms at the end of tenancies; some of these were sold to existing tenants. Other CUA, such as Gloucestershire, have adopted a more 'middle road' policy of deciding what is right for individual holdings at the end of each tenancy. This allows them to either use the land for the enlargement of existing county farms or to sell off small parcels of land for agricultural and non-agricultural uses.

Non-agricultural uses could help the CFE to satisfy its obligations to the wider rural community and to attract new entrants to farming. While rarely enacted during phases of growth and consolidation, different forms of diversification could represent a survival strategy for the CFE. Thus, Hampshire County Council has been keen to explore opportunities for the procurement of food produced by their tenants by local authority services, while Cambridgeshire has provided land and organised community tree planting events. Likewise, both Devon and Cornwall have developed educational centres on their CFE with a view to providing important facilities for school children and adding value to the county farm service. Such education services could be extended to incorporate environmental/conservation goals. These kinds of developments could lead to new markets for county farm tenants, the creation of local food processing jobs and increased educational awareness of the benefits of local food. CFE tenants could also diversify into different forms of farm-based tourism and recreation. Different combinations of agricultural and non-agricultural uses might prove attractive to younger and more innovative first time farmers, bringing vital new blood to the CFE.

While the concept of farm diversification is more likely to be accepted by new entrants than by some of the existing and longer-term CFE tenants, it must not be seen as contradictory to the original goals of allowing 'new blood' and non-successional tenants into farming and developing viable farming enterprises.

Indeed, recent policy debates in the UK have shifted attention away from local foods and diversification activities and towards more national and global issues associated with food security. In response to such factors as a rapidly rising world population, volatile energy and food prices, water resources and climate change, there is once again pressure to produce more food (Bridge and Johnson 2009, Chatham House 2009). This re-emphasis on agricultural productivism could have important implications for the future direction of the CFE and the attraction of 'new blood' into the industry, especially as the UK has a major deficit in the production of horticultural products, many of which could be grown on the smaller-scale county farms in different parts of the country. There is also potential for the CFE to link up more directly with initiatives like FS in Cornwall in order to attract new entrants to farming and the next section examines some of the developments relating specifically to the FS initiative in Cornwall.

The Fresh Start Initiative in Cornwall

Following the Commission for the Future of Farming and Food's recommendation to create new routes into farming (Defra 2002), the Fresh Start (FS) initiative was launched nationally in December 2004. In the early stages of the programme (March 2005), a pilot initiative was launched in Cornwall (Fresh Start in Cornwall) by the Cornwall Agricultural Council's Development Team (CACDT). It was intent on assisting Cornwall and the Isles of Sicily's farmers to rationalize, restructure and improve efficiencies through an integrated system of elements that were aimed at facilitating both entry into, and exit from, farming. Although it initially provided useful feedback as a pilot for the national FS scheme, the later scheme subsequently focused solely on new entrants and the setting up of FS Academies. The research presented here derives from an evaluation of the FS scheme in Cornwall between 2005 and 2008, and its attempt to simultaneously address both entry and exit issues within farming.

FS in Cornwall was an industry-led initiative, run by CACDT in partnership with Business Link and Duchy College and funded by the EU Objective One Structural Programme. It had three key aims: first, to encourage and help newcomers to the agricultural industry to ensure its long-term health and vitality; second, to encourage those within the industry to plan ahead and explore new opportunities and options; and third, and in contrast to the CFE, to provide an opportunity for those wishing to leave the industry to do so with dignity. In order to help achieve these aims, FS offered an innovative package of measures to applicants that were intended to make use of existing organisations and structures in the county, where possible. These included: matchmaking; business support; retirement and succession planning; mentoring; training; financial engineering; and rural housing provision. These were co-ordinated by a full time FS co-ordinator, but delivered by a range of partners using established networks and expertise. Key amongst these partners was the Cornwall CFE.

The matchmaking element was facilitated by the scheme's coordinator and involved trying to link new entrants to available holdings, or putting them in touch with existing farmers considering some kind of joint venture. Business support was available to those entering the industry on the basis of pre-tender, pre-investment and post-investment subsidised consultancy support, delivered through the existing mechanisms of the Rural Business Support Initiative (RBSI) and administered by Business Link Devon & Cornwall. Normally, this support is only available to those that are currently generating a minimum of £1500 income through farming and who have an agricultural holding number;[3] however, under FS eligibility was extended to include anyone who had been accepted onto the initiative. Equivalent support was also available for those leaving the industry, under the heading of 'retirement and succession planning'. In contrast, mentoring was envisaged as being distinctive from business support and intended to be primarily about setting goals and providing a listening ear. It also aimed to help build confidence, self-esteem and the ability to make decisions. It was anticipated that the mentoring would operate alongside and after business support had been withdrawn. In addition, and distinct from mentors, 'farm buddies' were introduced in 2006 to provide practical farming advice to new farmers such as giving advice on when to cut grass for hay, with up to a maximum of three days support.

In terms of training, Duchy College was allowed to spend the Voluntary Training Scheme (VTS) budget on 'eligible beneficiaries' within the FS initiative, including those who were not already in the industry. Duchy College offered a number of courses aimed at FS participants based on the results of training and business needs analyses that were conducted. Following some initial delays, the financial engineering element of FS was finally agreed with Government Office South West (GOSW) in April 2006. The approved budget was £250,000, with the intention of delivering eight loans at an average of £31,250. With respect to the affordable rural housing element, the intention was to work with organisations such as the Addington Trust to stimulate activity in the county and put some pressure on the local planning departments.

It can be seen, therefore, that FS provided a range of support to facilitate exit from, and entry into, farming. Nevertheless, in terms of hard outputs it has been limited; despite having 212 registrations of interested applicants, there were only six successful farm tenders and one successful joint-venture. In addition, business support was provided to 43 people, 14 received mentoring support, four farm buddy support, two VTS training, two retirement support and three people

3 An Agricultural Holding Number is assigned to land owners/managers who register their land with the government. This enables then to apply for the Single Farm Payment or other grant schemes from the Rural Payments Agency.

received loans through the financial engineering package. So, although numbers of applicants have been high, those who have gone on to train, acquire holdings or enter share farming arrangements with FS support have been low.

Matchmaking, as a means of facilitating some form of shared farming or joint venture, was acknowledged by stakeholders to be a good idea. However, in reality there proved to be a number of significant barriers to its success: new entrants' lack of capital, insufficient money being generated within the projected new business to support two families, and lack of housing for the new entrant. It also proved very difficult to match strangers together, and there was a perception that the effort might have been better spent facilitating the legalisation of agreements that had already been informally established, rather than trying to act as an initial matchmaker. Whilst the structural limitations to joint ventures have already been recognised (ADAS 2004), this evaluation has revealed that the human side of matching unknown parties also needs to be considered.

Business support was the most popular element with applicants, proving particularly valuable for new entrants applying for county farm tenancies. However, while a budget of £235,000 was made available by GOSW for business support, nearly 80 per cent of this budget remained unspent. It may be that the projected numbers submitted to GOSW were overly optimistic, but it also highlights the lack of opportunities available for new entrants trying to gain access to farms.

The uptake of the 'retirement and succession planning' initiative was minimal, mainly because farmers considering retirement do so over a long period of time, usually requiring professional legal and accountancy advice; as such, the support offered by FS was not seen as appropriate (or at least sufficient) to help them to make a decision to retire. Although the numbers receiving mentoring support were quite small, it was described as being significant for the small minority that did, with strong and supportive relationships being developed between mentors and those setting foot on the farming ladder for the first time. Only four FS clients used buddies, but had buddying been introduced from the beginning of FS it is likely that the uptake would have been better. Where the buddies were used, they were felt to have been helpful to the new entrants concerned.

In terms of training, the low uptake of VTS can be explained by the fact that those interested in training were thinking about learning short-term practical skills rather than enhancing their business and management skills. In this respect, FS has highlighted the need to provide support packages specifically targeted at industry restructuring, in addition to those that enable upskilling for those who are already in the industry. In a similar vein, only £84,200 out of the total budget of £250,000 for 'financial engineering' was allocated to three businesses. The intention was for this package to be used as a lender of last resort for those unable to get funding from normal lending sources, due to a lack of track record or collateral. Although some concerns were expressed that there was a danger of supporting businesses that banks would not consider, those running the package (South West Investment Group) emphasised that they based their decisions on the viability of the business case in much the same way as a commercial bank; the only difference being that a commercial bank would be looking for more tangible security. Furthermore, the fund enabled three people to get into farming who would not otherwise have done so. It was felt that had the fund been available earlier in the scheme three further new businesses might have been supported. It was also the case that when it started the interest rates offered within the financial engineering scheme were not very attractive; this situation changed as the scheme neared its conclusion.

Finally, there is little doubt that affordable rural housing is perhaps the key issue that needs to be addressed within the context of industry restructuring, but also more broadly within rural areas. However, this proved to be the most difficult for FS to engage with and the scheme failed to identify any suitable mechanisms for addressing this issue. A lack of suitable affordable housing has certainly acted as a block on mobility within Cornwall (which has exceptionally high property prices), most notably affecting older farmers' ability to move off their farm.

In theory, there is the potential for both the private and county farm sectors in Cornwall to provide tenancy opportunities for new entrants. However, in practice, both sectors are constrained by the need to retain a profitable tenanted farm structure on their estates. What this means is that when farms do become available, there is usually demand from existing tenants within the estate to increase the size of their holdings. Consequently, the size of holdings has tended to increase beyond the size appropriate/affordable for a first time entrant to farming. Although the CFE supported FS (indeed the only successful FS new entrant tenancies were with county farms), there was increasing frustration amongst applicants that county farms are not sympathetic to younger people and that their land agents are under pressure to make the CFE economically viable in the face of a range of budgetary pressures. It is also clear that land agents on the private estates face similar pressures. Yet ironically, representatives of both the private estates and county farms acknowledge that they need to reinvigorate their farm estates with young farmers. In this respect, the evaluation suggested that schemes like FS have a potentially 'linking role' in facilitating young farmers' entry into the county farm estate, before then progressing them onto private estates. Nevertheless, in practice there were no examples of this happening.

Given the competition for county farm tenancies and the limited opportunities on private estates, many participants in the evaluation felt that land being released from retiring land owners offered the best hope for new entrants. However, barriers were identified here as well. Farmers thinking of retiring or winding down were found to be often looking for easy and/or short-term solutions, as well as considering their taxation position. For them, dividing up the farm and buildings and letting out or selling their land (often to neighbours) is likely to be the easiest option, meaning whole farms do not often become available on the open market. This tendency towards taking a short-term 'holding position' was reinforced by uncertainties associated with the SPS in the early years of the scheme. Furthermore, this approach enables farmers to stay in their homes, an understandable consideration given the shortage of affordable rural housing in the region. These considerations, which have been articulated by other researchers (e.g. Williams 2006), can help explain, at least in part, why FS has failed to engage with the retirement community. However, there was also a perception amongst the evaluation participants that 'older' farmers were often not aware of what FS might have to offer them. FS applicants in particular felt that FS should have been more proactive in linking those who want land with those that may be considering retirement. The consultants interviewed also stressed that the focus of FS should have been on making it easier for people to exit farming, in that this was recognised as the principal 'bottleneck' to enabling 'new blood' to come into the farming industry.

In terms of facilitating non-successional entry into UK farming, FS has clearly been largely unsuccessful. Nevertheless, it is important to reflect that the FS scheme has been operating against the backdrop of a lack of available land, a lack of finance/capital available to new entrants and a lack of affordable rural housing (a particular problem within Cornwall), all of which have restricted the opportunities for both new entrants and retiring farmers. The net effect of this has been that the number of holdings coming onto the market has been insufficient to meet the needs of aspiring new entrants in the region, causing a 'bottleneck' in terms of the scheme's operation and effectiveness – notwithstanding that it was this blockage that FS was set up to address. In essence, FS as a policy intervention has been unable to overcome the structural barriers that new entrants face or, as one consultant pointed out, 'you can't buck the market' however well-intentioned your initiative is. Nevertheless, there was recognition by stakeholders and applicants alike that FS had been worth a try. It represented an innovative approach to the entry-exit issue and provided some useful lessons for future policy interventions, as discussed.

Discussion and Conclusions

Problems of entry to and exit from farming in the UK are well-documented, a combination of a reduced rate of entry by new, young farmers and a reduced rate of retirement by older farmers (ADAS et al. 2004). Family succession, as the main route into farming, has dominated studies of agricultural adjustment and the role of succession. These studies – like most other chapters in this book – are about understanding the transfer of farm business assets between generations. The material presented in this chapter has *not* been about 'keeping it in the family', examining instead public policy initiated interventions – the CFE and the pilot FS initiative in Cornwall respectively – that have been designed to enable 'new blood' to enter the farming industry. The analysis shows that, although non-successional entrants into farming face a number of distinct challenges such as a lack of capital and land, they are also subject to the same wider structural forces that affect family farm succession.

The two schemes have contrasting geographical reach and start-up periods. The CFE is a well-established national scheme and has historically acted as an important first rung on the farming ladder. The historical and geographical analysis presented here shows that, despite an impressive growth in the number of farms in the first half of the twentieth century, the CFE has more recently experienced a decline in farm numbers through processes of farm amalgamation and disposal by county and unitary authorities. The CFE currently accounts for only 1 per cent of the agricultural area in England and Wales, with a notable reduction in dairy farms, as well as a fall in the provision of equipped holdings. The CFE equates to only 3 per cent of tenanted land in England, although significantly it was responsible for just over one-third of all new let, fully equipped holdings in 2006.

Evaluation of the more recently established FS pilot in Cornwall is equally gloomy; in essence, it failed to facilitate non-successional entry into farming, with limited uptake across a range of measures, which included matchmaking, business support and retirement, and succession planning. For example, despite 212 registrations of interested applicants, there were just six successful farm tenders, one successful joint-venture and two instances of retirement support being provided.

It is clear in both cases that wider structural and market pressures have significantly influenced the nature, role and success of the schemes. No longer obliged to provide smallholdings, a number of CUA are increasingly pressured and inclined to sell off their farm estates to benefit from high land prices. Equally, the success of the FS initiative in Cornwall has been limited by a lack of holdings because of competition from the public and private tenanted sector, as well as by pressures on capital and the rural housing market.

The demise of the CFE has not gone unnoticed. A recent report by Sir Donald Curry (2008) argues, for instance, that the County Council Farm structure needs to be retained as one of the few entry points for those wanting to start a business. The CFE is rightly championed as an indispensable national asset for the structure of UK agriculture. This same report also notes the service benefits that the estate can bring to local authorities in meeting wider objectives in relation to countryside and environmental issues, including, e.g. renewable energy, waste utilisation, local food, access to the countryside, learning outside the classroom, planning policies, greenbelt management and assisting in the management of flood risk. In short, the CFE, and its espousal of 'service provision', embodies, if managed properly, the multi-functional properties of modern farming. However, as is clear from the analysis here (see also Ilbery et al. 2010a), a more strategic approach is necessary to realise these potential benefits; when properly utilised (as is the case in some local authority contexts), such farms become significant nodes of learning and demonstration, enabling new ideas, as well as new people, to enter farming and demonstrate new change processes.

Both schemes have been influenced by the effect of stagnation. In the case of the CFE, only a small majority of farm tenants have taken up farm opportunities elsewhere. They have not enabled new farmers to succeed them, a problem not helped by the relatively high average age of tenants, effectively mirroring the succession barriers noted in the family farming literature (see Brookfield and

Parsons 2007). This stagnation would appear to be the consequence of poor management, or at least a lack of management relative to the aims of the scheme. Smaller county council farms arguably should only ever be viewed as starter units (as originally intended), and local authorities could use FBTs as a way to ensure that prospective tenants who are taken on understand that they should use the tenancy as an opportunity to establish a business and to then move on to a larger unit within a specified timescale (Curry 2008), thus allowing others to succeed them. In short, tenancies must be managed and succession plans established, thereby removing the temptation to remain on a farm and not move on, so often the primary barrier in family farm contexts. In the case of the FS analysis, stagnation resulted largely from the farmers' ability and desire to stay on the farm, accompanied by their inability to afford alternative housing for their retirement. This can, in part, explain the lack of FS's engagement with those considering retirement; however, the inappropriate nature of the support offered was also identified as an issue.

The impact of such government-initiated schemes is clearly significantly determined by wider processes of structural change. This relationship with broader structural processes is also seen in mainstream farm succession work (Williams and Farrington 2006), with farm household intentions influenced by local context, especially surrounding labour markets. Research by ADAS et al. (2004) has shown, for instance, that the low rate of new entry into family farms reflected, at that time, the poor returns compared to alternative occupations rather than any barriers to entry. In a survey of students of land-based subjects, they found that they did not intend to run a farm business by the time they leave college, mainly because of the positive attractions of other careers, although most intended to come back to run a farm in the future. In other words, entry and exit rates are flexible and will be influenced by profitability and stability within the industry.

The current farming environment has clearly changed and there is now a stronger case to be made about considering a career in farming. The nature of the CFE still remains challenging, especially as a number of local authorities come under increased pressure to reduce spending and maximise assets. Nevertheless, given food and energy security demands, coupled with pressures to maintain and protect environmental services and assets (Lobley and Winter 2009), the demand and need for initiatives that have the potential to enable new farm families to enter the industry remain arguably stronger now than it has done in the past. Non-successional schemes that enable new blood to enter any industry are important to improve and fill skills shortages, encourage entrepreneurship and new ideas, improve resilience and adaptability, and, in the case of farming, to help reduce an ageing working population. The National Fresh Start initiative, in the form of its National Stakeholder Group, has recognised this need to 'mend the farming ladder' and recently launched a campaign to meet with large private and institutional landowners to better enable new entrants to access the industry. Similar actions are urgently required on the CFE side, to ensure it maintains its estate portfolio and continues to complement and support traditional family succession pathways. However, as this analysis as shown, any new initiatives to support new entrants must be accompanied by interventions to encourage and help older farmers to move on or leave the industry.

References

ACES (Association of Chief Estates Surveyors and Property Managers) 2007. *ACES Rural Practice Branch Rationale* [online]. Available at: http://www.aces.org.uk/rural/rural_about_ACES_rural.php4 [accessed 27 November 2007].

ADAS Consulting Ltd, University of Plymouth, Queen's University Belfast, Scottish Agricultural College 2004. *Entry to and Exit from Farming in the United Kingdom*.

Bridge, J. and Johnson, N. (eds) 2009. *Feeding Britain*. The Smith Institute: London.

Brookfield, H. and Parsons, H. 2007. *Family Farms: Survival and Prospects. A World-wide Analysis*. Routledge: Abingdon and New York.

CAAV (Central Association of Agricultural Valuers) 2007. *Annual Tenanted Farms Survey 2006*. Central Association of Agricultural Valuers: Coleford.

Cannadine, D. 1990. *The Decline and Fall of the British Aristocracy*. Yale University Press: London.

Chatham House 2009. *Food Futures: Rethinking UK Strategy*. A Chatham House Report. Chatham House: London.

CIPFA (Chartered Institute of Public Finance and Accountancy) 2007. *County Farms and Rural Estates Statistics*. The Chartered Institute of Public Finance and Accountancy: London.

Curry, D. 2008. *The Importance of the County Farms Service to the Rural Economy*. Prepared by the Sir Donald Curry in collaboration with the Tenant Farmers Association, the Country Land and Business Association, the Royal Institution of Chartered Surveyors and the National Farmers' Union [online]. Available at: www.Defra.gov.uk/farm/policy/sustain/pdf/county-council-farms.pdf [accessed 25 May 2010].

Defra 2002. *Farming and Food: A Sustainable Future*. Policy Commission on Farming and Food: London.

Defra 2006. *Agriculture in the UK 2006*.

Drummond, I., Campbell, H., Lawrence, G. and Symes, D. 2000. Contingent or structural crisis in British agriculture? *Sociologia Ruralis*, 40, 111–127.

Errington, A., Millard, N. and Whitehead, I. 1998. *The Emerging Shape of Farm Business Tenancies*. Department of Land Use and Rural Management, University of Plymouth.

Ilbery, B. and Maye, D. 2008. Farm tenancy. *Geography Review*, 21, 32–35.

Ilbery, B., Ingram, J., Kirwan, J., Maye, D. and Prince, N. 2010a. Structural change and new entrants in UK agriculture: examining the role of county farms and the Fresh Start initiative in Cornwall. *Journal of the Royal Agricultural Society of England*, 170, 77–83.

Ilbery, B., Maye, D., Watts, D. and Holloway, L. 2010b. Property matters: agricultural restructuring and changing landlord-tenant relationships in England. *Geoforum*, 41, 423–434.

Lobley, M. and Potter, C. 2004. Agricultural change and restructuring: recent evidence from a survey of agricultural households in England. *Journal of Rural Studies*, 20, 499–510.

Lobley, M. and Winter, M. 2009. Introduction: knowing the land. In *What is Land For? The Food, Fuel and Climate Change Debate*, edited by M. Winter and M. Lobley. Earthscan: London, 1–20.

Newby, H., Bell, C., Rose, D. and Saunders, P. 1978. *Property, Paternalism and Power*. Hutchinson: London.

NFU (National Farmers Union) 2007. *News Extra* [online]. Available at: http://www.countrysideonline.co.uk/news.php?extend.2327 [accessed 6 November 2007].

Northfield Report 1979. *Report of the Committee of Enquiry into the Acquisition and Occupancy of Agricultural Land.* Her Majesty's Stationery Office: London.

Smith, N. 1946. *Land for the Small Man: English and Welsh Experience with Publicly-supplied Small Holdings, 1860–1937*. King's Crown Press: New York.

Whitehead, I., Errington, A., Millard, N. and Felton, T. 2002. *An Economic Evaluation of the Agricultural Tenancies Act 1995*. The University of Plymouth: Newton Abbot.

Whitehead, I. and Millard, N. 2000. *Contemporary Issues of County Farm Estates in England and Wales*, prepared for RICS.

Williams, F. 2006. Barriers facing new entrants to farming – an emphasis on policy. Royal Geographical Society – IBG Annual Conference. London, 1–10.

Williams, F. and Farrington, J. 2006. *Succession and the Future of Farming: Problem or Perception?* Paper presented to The Rural Citizen: Governance, Culture and Well-being in the 21st Century Conference, University of Plymouth, March 2006.

Winter, M. 1996. *Rural Politics*. Routledge: London.

Wise Committee Report 1966. *Departmental Committee of Enquiry into Statutory Smallholdings*. Her Majesty's Stationery Office: London.

Chapter 8
So What?

John R. Baker

Introduction

The title of this chapter takes the form of a question: 'So what?' So what do we know about farm business succession? So what research is available on-farm business succession planning? So what does the research tell us about farm family business succession? So what have we done with the research findings? So what planning for a business succession are farmers doing? So what are the educational materials that have been developed to address the issues identified? So what policy changes have been made to address the issues identified by the research? So what … ? The number of 'So what' questions is probably quite large and in this chapter only a few are addressed.

This chapter is based on two main sources: the results of *FARMTRANSFERS* research and forty years of experience, including twenty-four years as a practicing attorney and founder and administrator of the Beginning Farmer Center located at Iowa State University.

In the introduction we noted that several of the chapters in this book are based upon the research and findings of the *FARMTRANSFERS* international research project. This chapter will report selected findings from the international research and selected findings from three states in the United States. Lastly, the chapter will describe educational techniques that address several issues identified by the *FARMTRANSFERS* project.

Some International Comparisons

Three findings from the comparison of the international data are of particular interest. These are the decisions concerning retirement, with whom retirement has been discussed and what decision making authority has been retained by the owner or transferred to the successor.

The greatest percentage of those responding to the survey indicated they would either never retire or semi-retire. The only exception was France where nearly two thirds of those surveyed responded that they would retire. It must be noted that at the time of the French replication of the survey a mandatory retirement scheme was in place. With the exception of France, the most frequent response was semi-retirement. Indeed in Australia and England over half of respondents stated that they would semi-retire (see Table 8.1).

When asked with whom retirement had been discussed, the most frequent response by United States respondents was either family or no one. Indeed, in North Carolina and in the 2000 Iowa replication the number one response was that retirement had not been discussed with anyone.

The last international comparison is the transfer of decision making authority from the owner to the successor. The rank order of retained authority is set forth in Table 8.3. It is interesting that in every country listed, the last two decisions that are transferred to the successor are deciding when to pay bills and negotiating loans and financing.

Table 8.1 Farmers retirement plans (per cent of respondents)

	Aus.	England	France	Ontario	Virginia	North Carolina	New Jersey	Penn.	Iowa 2000	Iowa 2006
Will Retire	32	34	65	34	24	18	24	24	27	23
Semi-Retire	56	53	29	44	34	35	29	29	38	47
Never Retire	11	13	6	22	42	47	40	40	35	31

Source: *FARMTRANSFERS*

Table 8.2 Discussion of retirement plans

	Aus.	Eng.	France	Ontario	Virginia	North Carolina	New Jersey	Penn.	Iowa 2000	Iowa 2006
Family	59	28	55	63	66	25	31	31	46	53
Lawyer	9	14	0	7	10	0	9	9	17	18
Banker	10	7	9	10	1	0	3	3	8	14
Accountant	40	39	39	38	11	0	13	13	19	10
Farm	8	4	21	5	2	0	2	2	3	4
Consultant	9	0	8	0		0	0	0	0	0
Other farm Advisor	5	7	6	7	0	7	4	4	0	7
Other	9	44	28	10	1	1	2	2	3	4
No one				28	30	67	21	21	47	36

Source: *FARMTRANSFERS*

Table 8.3 Delegation of decision making (rank order) in selected *FARMTRANSFERS* surveys

Activity/ Decision	Iowa 2006	Iowa 2000	Aus. 2004	Eng. 1997	France 1993	Ontario 1997	Quebec 1997	Virginia 2001	Japan 2001
Decides when to pay bills	2	1	1	1	1	1	1	1	2
Identify sources and negotiate loans and finance	1	3	2	2	3	2	2	2	1
Decide long-term balance and type of enterprises	12	7=	3	6	6	7	10	5=	11
Decide and plan capital projects	3=	4	4	5	7	5	8	7	9
Negotiate purchase of machines and equipment	8	5	5	8	9	6	9	8	12
Decide when to sell crops/ livestock	3=	7=	6	4	4	4	5	5=	6=
Negotiate sales of crops/ livestock	3=	2	7=	3	2	3	3	4	6=
Make annual crop/ livestock plans	7	10	7=	7	5	8	4	9	4
Level of inputs used	6	6	8	13	13	11	6	3	5
Plan day-to-day work	10	11	9	9	8	12	11	12	3
Decide timing of operations/ activities	11	12	10	10	12	9	7	10=	8
Decide type and make of machines and equipment	13	9	11	11	11	10	12	10=	13
Decide work method/way jobs are done	9	13	12	12	10	13	13	13	10

Source: *FARMTRANSFERS*

So what are the implications of these results? One may be that when the owner does not retire there may not be room in the farm business for the successor. A second implication is that when the owner semi-retires, decisions with respect to financial matters are not transferred to the successor. Lastly, it appears that in many states and countries, farm owners are not soliciting or seeking professional advice concerning retirement. In some cases this may be due to the fact that they are not considering retirement and therefore do not see the need of such advice.

Delegating Decision Making in Three US States

Of particular interest is the delegation of decision making authority from the owner/operator to the successor and which factors, if any, affect such delegation. Three very different states have been selected for comparison. The states are Iowa, Pennsylvania and New Jersey. The states were chosen because data was available and because of the differences in the area, number of farms, size of farms and size of population.

Iowa has a population of approximately three million (2010 US Census). According to the 2007 Census of Agriculture, there were 92,856 farms and the average size farm was 331 acres. Iowa encompasses approximately 37,000,000 acres of which approximately 26,300,000 are used in agricultural production. The population of New Jersey is approximately 8.7 million (2010 US Census). According to the 2007 Census of Agriculture, there were 10,327 farms and the average size farm was 71 acres. New Jersey encompasses approximately 5,400,000 acres of which approximately 500,000 are used in agricultural production.

The population of Pennsylvania is approximately 29.5 million (2010 US Census). According to the 2007 Census of Agriculture, there were 63,103 farms and the average size farm was 124 acres. Pennsylvania encompasses approximately 29,500,000 acres of which approximately 4,900,000 are used in agriculture.

From Table 8.4 it is clear that Iowa operators are more likely to delegate decision making authority than are their counterparts in New Jersey and Pennsylvania. This may be a function of size of operation that mandates that decision be shared and delegated. As we have seen, farms in Iowa are considerably larger. In addition, it may be that along with size, the complexity of the farm enterprises may require the sharing or delegation of decision making. Although *FARMTRANSFERS* data can help identify such differences, further research is required in order to understand why they occur.

Table 8.4 Delegation of decision making authority by farm operators in Iowa, New Jersey and Pennsylvania

Iowa	**Decision made by:**				
	Operator alone	**Shared between the operator and the successor**			**Successor alone**
	1	**2**	**3**	**4**	**5**
Decide when to pay bills	44%	20%	19%	3%	14%
Identify sources and negotiate loans and financing	47%	21%	19%	2%	11%
Plan day-to-day work	18%	31%	32%	6%	12%
Make annual crop/livestock plans	19%	29%	37%	5%	10%
Decide the mix and type of enterprise in the long run	16%	31%	34%	8%	11%
Decide the level of inputs to use	25%	25%	33%	5%	13%
Decide the timing of operations	15%	30%	39%	3%	12%
Decide when to sell crop/livestock	27%	27%	31%	6%	8%
Negotiate sales of crops/livestock	31%	21%	35%	3%	10%
Decide type and make of machinery and equipment	16%	30%	34%	11%	9%
Negotiate purchase of machinery and equipment	23%	27%	31%	6%	13%
Decide when to hire more help	21%	33%	28%	9%	9%
Recruit and select employees	24%	24%	30%	6%	16%
Decide amount and quality of work	24%	29%	33%	5%	9%
Supervise employees	25%	29%	31%	6%	8%
Decide work method/way jobs are done	18%	26%	41%	7%	8%
Decide and plan capital projects	24%	32%	29%	8%	6%
Livestock management	19%	20%	35%	9%	17%
Keeping farm records	45%	16%	19%	5%	15%
Average 2006	25%	26%	31%	6%	11%
Average 2000	50%	20%	13%	6%	10%

New Jersey	**Decision made by:**				
	Operator alone	**Shared between the operator and the successor**			**Successor alone**
	1	**2**	**3**	**4**	**5**
Decide when to pay bills	67%	16%	9%	2%	6%
Identify sources and negotiate loans and financing	65%	18%	10%	3%	4%
Plan day-to-day work	48%	26%	15%	4%	7%
Make annual crop/livestock plans	47%	31%	14%	4%	3.5%
Decide the mix and type of enterprises in the long run	37%	31%	21%	6%	5%
Decide the level of inputs to use	43%	34%	13%	3%	7%
Decide timing of operations	48%	26%	15%	2%	9%
Decide when to sell crop/livestock	46%	27%	15%	2.5%	8.5%
Negotiate sales of crops/livestock	50%	26%	14%	1%	9%
Decide type and make of machinery and equipment	46%	28%	16.5%	0%	9%
Negotiate purchase of machinery and equipment	51%	25%	13%	1%	10%

Decide when to hire more help	52%	26%	12%	2%	8%
Recruit and select employees	51%	22%	12%	1%	13%
Decide amount and quality of work	43%	30%	16%	2%	9%
Supervise employees	41%	29%	13%	4%	13%
Decide work method/way jobs are done	39%	30%	19%	2%	10%
Decide and plan capital projects	47%	32%	13%	3%	5%
Livestock management	37%	29%	15%	4%	15%
Keeping farm records	59%	22%	9%	1.5%	8%

Pennsylvania	**Decision made by:**				
	Operator alone	**Shared between the operator and the successor**			**Successor alone**
	1	**2**	**3**	**4**	**5**
Decide when to pay bills	69%	15%	7%	2%	7%
Identify sources and negotiate loans and financing	65%	20%	6%	3%	5%
Plan day-to-day work	44%	27%	17%	4%	8%
Make annual crop/livestock plans	40%	34%	16%	5%	5%
Decide the mix and type of enterprises in the long run	37%	31%	22%	5%	6%
Decide the level of inputs to use	44%	30%	16%	3%	7%
Decide timing of operations	48%	25%	15%	4%	8%
Decide when to sell crop/livestock	47%	26%	17%	3%	8%
Negotiate sales of crops/livestock	50%	24%	15%	3%	8%
Decide type and make of machinery and equipment	43%	29%	18%	3%	7%
Negotiate purchase of machinery and equipment	51%	23%	15%	3%	8%
Decide when to hire more help	53%	21%	13%	2%	10%
Recruit and select employees	50%	21%	14%	3%	11%
Decide amount and quality of work	44%	28%	16%	5%	7%
Supervise employees	44%	27%	13%	5%	11%
Decide work method/way jobs are done	37%	34%	18%	4%	7%
Decide and plan capital projects	48%	28%	15%	3%	5%
Livestock management	37%	27%	18%	5%	12%
Keeping farm records	59%	22%	8%	3%	8%

Source: FARMTRANSFERS

Communication

Communication is the indispensable skill for farm business succession planning. Every aspect of farm business succession planning requires that everyone involved be able to communicate effectively. Communication is more than talking; it requires listening. Communication is required in management, planning, developing and maintaining relationships, directing others, teaching others, learning from others; it is an essential skill in our lives. It is difficult to imagine any activity, business or non-business, which does not require communication.

Individuals are made up of a variety of attributes that influence how effectively that individual communicates. In developing a farm business succession plan three attributes play an important, if not dominant role. These attributes are gender, generational placement and personal evolution.

No one would be surprised by the fact that gender has a significant impact on how an individual communicates, and that men and women tend to communicate in different ways. Whether this difference is inherent or acculturated is immaterial to the process of farm business succession planning. What is important is the recognition and acceptance that gender based differences exist and that these differences influence how we communicate.

Deborah Tannen in her book, *You Just Don't Understand: Women and Men in Conversation* (1992), listed some of the differences in men's and women's communication. She noted that men tend to communicate to establish status and demonstrate their independence. They give advice and seek information. They are comfortable in giving orders and being involved in a conflict. Women, on the other hand, tend to communicate to receive support and establish intimacy and understanding. They are concerned with feelings and rather than give orders will make proposals that may lead to a compromise.

The generation into which we were born and the events we experience affects how we communicate. It is not uncommon for three generations, and often times four generations, to be involved in farm family business succession planning. This occurs when the grandparents own the farm, an heir manages the farm and a grandchild or great-grandchild has expressed a desire to join the farm family business. These generations have been labelled, 'the Silent Generation', 'the Baby Boomers', 'generation X' and, lastly, 'generation Y', respectively. The events and markers which these generations have experienced have influenced their values (University of Iowa School of Social Work 2009):

- The generation born between 1925 and 1945 is 'the Silent Generation'. The events and markers of this generation include the great depression, World War II and the Korean War. The values of 'the Silent Generation' include loyalty, respect for others and hard work and sacrifice (University of Iowa School of Social Work 2009).
- The generation born between 1946 and 1964 is 'the Baby Boom Generation'. The markers of this generation include the civil rights movement, the Viet Nam War and the women's liberation movement. The values of 'the Baby Boomer Generation' include optimism, discovery and personal fulfilment (University of Iowa School of Social Work 2009).
- The generation born between 1965 and 1980 is 'Generation X'. The markers of this generation include the Watergate break-in, both parents working and the emergence of MTV (music television). The values of 'Generation X' include balance, self-dependence and embracing diversity. (University of Iowa School of Social Work 2009).
- The generation born between 1981 and 2000 is 'Generation Y'. The markers of this generation include increased racial diversity, advances in technology and changes in education. The values of 'Generation Y' include

> a global orientation, interest in health, fitness and body image and a focus on technology (University of Iowa School of Social Work 2009).

The third attribute that influences communication is personal experience. Our experiences influence our development and it is true that we change as our experiences change. Parents will not be surprised by the statement that children are different from the day they are born. We are all 'hard wired' with certain innate interests and abilities. Our innate interests and abilities lead us to certain activities and actions that we enjoy doing. We can all remember the childhood friend who could draw and who spent most of his or her time drawing. We may know someone who is a natural musician and can play a variety of instruments. Because we enjoy an interest, we engage in the activity associated with the interest; in other words we practice the activity, and the more we practice the better we become. With increased proficiency we seek opportunities to engage in the activity and the activity is the focus of our attention and efforts.

When the attributes of gender, generation and personal evolution are recognized and appreciated, communication is enhanced. The recognition and appreciation of these differences in styles, values and interest will mitigate conflicts.

Communication involves listening as well as talking. When we listen in a conversation we will listen for the facts that are being communicated. We will also hear the emotion, the feelings associated with the facts. Consider the following scenario:

> The owner generation is involved in the development of a farm business succession plan. It is a fact that the farm business will transition to a succeeding generation. It is also a fact that the owner is giving up control of the farm and is, in fact, quite frightened by that fact.

Should the successor listen for the fact that the farm will be transferred or for the fact that the owner is frightened? Listening for fact is important but listening for feeling, for the emotional content, is equally important. The same fact may elicit different emotional responses from different people. To understand and appreciate what others are communicating, we must listen for the emotions associated with the facts with the same attention that we listen for the facts.

How we understand what is being communicated, the perception of what is being communicated is equally important. The expressions 'taking it the wrong way' or 'that's not what I meant' convey the problem associated with how communication is perceived. The two expressions illustrate instances where the listener perceived a message different than was intended.

Effective communication is a skill that can be learned and with practice can be improved. The starting point to effective communication is the awareness and appreciation of the role that gender, generation and personal play in communication is. Listening carefully for the fact, emotions and how the communication is perceived enhance effective communications. A few simple techniques will also improve communication.

Pay attention to the person who is speaking. Look at them and listen to what they are saying. Listen for the facts, the emotions associated with the facts, and monitor how you perceive what is being said. Summarize and allow the speaker to correct any misperceptions that you may have. Ask questions to discover additional information or to insure an accurate understanding of what is being said. If you are speaking invite a response. Ask what others 'think' and how they 'feel' about what you have said. If you are asked a question respond to it appropriately. Clarify any statements that are ambiguous. Provide additional facts or express your feelings in an accurate or complete manner.

Avoid behaviour that blocks effective communication, the behaviour or statements that inhibit communication. Behaviours and statements that inhibit communication include giving orders or threatening and warning of the consequences of behaviour. Moralizing, lecturing and proffering unsolicited and unwanted advice will also adversely affect communication.

What's Important?

'Do I want to pass my farm business to a successor or do I want to pass a group of farm assets to my heirs?' On what basis will the answer be made to this initial question? And if the decision is to pass the farm business to the successor what will be the basis for all the subsequent decisions? This is the central question that must be asked and answered by the owner generation. If it is their desire to pass a group of assets to their heirs, the 'keeping the land in the family' priority, a simple estate plan will accomplish that. Throughout the United States and most other countries whose law is based upon English common law, if the owner dies without a will, that is the owner is intestate, the law will mandate that the property pass to a spouse or to the descendants of the owner. If, however, the owner desires to pass the farm business intact to a successor then income, some assets and management must be transferred from the owner generation to the successor.

When a decision must be made the decision maker will select the alternative that will result in the wanted outcome. In a general sense the wanted outcome is important to the decision maker and it is important because it conforms to the decision maker's values and value system. Simply stated, that which is important is that which is valued. It is therefore logical that a farm business succession plan should be an expression of the values of those involved in and affected by the plan.

People will not do anything that they truly do not want to do. However, people very often do things they do not like to do, in order to accomplish that which they want to do. All too often the farm business succession planning process begins with goal setting. Goal setting is indeed an important step in developing a succession plan, but it cannot be the starting point, as the goals are set to accomplish the mission, the mission is how the envisioned future will become reality and the vision is an expression of what is important and that which is important is that which is valued.

The business succession planning process developed by the Beginning Farmer Center is based upon the following activities:

Identification of Values: What is important to me? What do I hold dear?
Creating a Vision: What do I want in the future?
Determing the Mission: How will this business create that future?
Setting the Goals: What do we want to do and what do we want to be?
Establishing Objectives: How will we measure our progress in achieving our goals?
Developing Strategies: What is the plan for accomplishing our goals?
Enlisting Tactics: What do we need to do to achieve our goals?

One method of identifying values is to generate a list of words that address the following:

What do you like or dislike in other people? Why?
Who are your heroes?
Of whom are you proud?
What do you consider your failures?
What do you consider your successes?
What are your fears?
Are there products you buy or do not buy because of advertising? Why?
What do you purchase?
Where do you purchase?
What do you hold sacred?
To whom are you loyal?
Who is loyal to you?
What factors do you considered when making decisions?
List ten tangible items that you want to own.
List ten attributes of character that you admire.
What charities do you support?
What do you like to do?
What don't you like to do?
What else do you value?

Not all values are of equal importance and it is necessary to prioritize the list, with the most important at the top of the list and the least important on the bottom. The succession plan must recognize and incorporate the values of those involved. The identification and enumeration of the values of the individuals, the family and the farm business are expressed in the mutually agreed upon vision of the future of the individuals, the family and the farm business.

An Intention

The old saw, 'If you don't know where you are going, any road will take you there' is certainly true in farm business succession planning. Another old saw that is equally true is that if you fail to plan, you plan to fail. Farm business succession planning is an outcome based process in that one must first decide where they want to be and when they want to get there. Once the 'where and when' have been established the 'how' can be determined. The 'where and when' are set forth in the vision statement that includes the values of the individual, the family and the farm business.

A vision statement describes where you want to go and what it will look like when you arrive. It is not a dream. In a dream, even in a daydream we can ignore reality and assume away impossibilities. In a vision statement we accept reality and recognize limitations and impossibilities. If the vision statement is to be an accurate description of the future that is desired and is to motivate all involved to work to achieve it, it must be a future that is achievable and important, that is valued by all. A vision statement provides a realistic picture of a desired destination. It is the where and when for ourselves and our farm business. It captures an accurate and realistic description of the surroundings and how the individual and the farm business acts within and reacts to the surroundings. It should inspire the family and the business to undertake the necessary activities to make the envisioned future a reality.

Creating a vision statement for the farm family business begins with the creation of a vision statement for the individuals involved in the farm business. It is the conflation of the individuals' vision statements that will become the vision for the business. When providing guidance to farmers at the Beginning Farmer Center we explain that:

> An individual vision statement should describe with particularity and specificity the future that is envisioned and your role in that future. It should describe in detail what you are doing and why you are doing it. Also include a description of your areas of responsibility, you authority and to whom you report. Include who with you, what they are doing and why they are doing it. Describe in detail your personal life, the amount of time devoted to family, friends and avocation and your social interactions.

To be useful, a statement must be shared and mutually agreed upon. If there is a conflict concerning the future of the farm business, and the individual's role in that future, the likelihood that the desired future will be created is greatly diminished. As the vision statement for the farm business is conflated from the individual vision statements it is important to pay particular attention to where and when the visions are congruent, when they are divergent and when they are in conflict. The conflict must be resolved and divergences reconciled in the farm business vision statement.

The mission statement sets forth why and how the farm family business will create the envision future. It should capture the individual's and family's values and vision. The mission statement is a sieve thought which all farm business decisions may be passed. In answers the questions: Does the decision aid in accomplishing the mission and If does not, why are we considering it? A mission statement communicates to customers and to the community why the business exists and helps to focus the owners on the essential activities that must be completed to create the desired future. It is extremely important the mission statement be a short and plain statement that is easily remembered and meaningful to those involved operating and managing the farm family business.

The mission statement should address the following question of why does the farm exist and what makes it unique:

> Consider how you want your neighbours, customers, suppliers and community to describe your farm business. Is the farm family business responsible to anyone other than the family? What need is filled by the farm business how does it fill that need? What makes your farm and the products it produces unique and why would anyone want those products or want to do business with your farm family business?

It is difficult to write a mission statement that does not become so long as to make difficult, if not impossible, to remember thereby rendering it useless as the guiding principle of the business. It would be rare indeed to create a meaningful and memorable mission statement in a single try. And as the family grows and changes, the farm family business will grow and change too. Assets will be bought and sold, new production techniques will be adopted and old techniques discontinued, new enterprise will be initiated and old enterprises discontinued. These changes require the mission of the farm to be revisited and rewritten when necessary.

A statement of intent combines the vision statement, the description of where and when, with the mission statement, the how, into one overarching statement. While most farm families grasp the importance of creating a vision statement and a mission statement they often have difficulty in creating a statement of their intentions. One teaching technique that has proven to be successful is to change the context to a future date. For example:

> Assume that it is 10 years in the future and you are writing the annual Christmas letter to your friends. As you write the letter consider what is happing in the industry, in your community, in your life and on your farm. You have decided to write in great detail about you, your family and your farm. In the letter you will write about what you are doing and where you are living. You will write about your family and what each family member is doing and where they are living. You will also write about the farm and include a description of the farm business assets, what, where and how much of each asset. You will include an explanation

> of why the farm has the asset and how it has helped the farm business. Include a description of who is involved in the farm family business, what they are doing and why they are doing it. Describe the financial condition of the farm and state in realistic terms the profit the farm is generating. You will write about the decision that you made and the reason you made those decisions.

When the exercise is complete the letter can be edited and condensed into the statement of intent. The statement of intent should provide the description of the future and how the farm family business will take the family to that future.

What's it Worth to Stay on the Farm?

> I've worked my entire life to pay off my uncles. Now I'll have to work the rest of my life to pay off my brothers (UK farmer, 2001).

This comment was made by an English farmer after hearing his father say that he intended to leave the farm in equal shares to his three sons. Unfortunately for heirs working in the farm family business the father's plan for the future of the farm is not uncommon. In the 2006 Iowa replication of the *FARMTRANSFERS* survey (Baker and Epley 2009), farmers were asked if they had made an Estate Plan. A total of 467 respondents indicated that they had indeed made an estate plan. As Figure 8.1 indicates, the respondents clearly intended for the farm land to stay in the family, with approximately 70 per cent using either wills, trusts, inter vivos gifts or sale to a family member.

Farmers with more than one potential heir make different provisions in their Estate Plans. Some plan to keep the farm as one unit and pass it on to one heir; others divide the property equally between all heirs. When asked what the best estate plan was, over 40 per cent of the respondents indicated their intention to divide the farm equally amongst all heirs, and less than 20 per cent intend for the successor to inherit the farm or a larger portion of the farm than other heirs (see Figure 8.2). A majority of those who responded that it was their intention to divide the farm business assets equally among all heirs stated that an equal division is 'fair'. Setting aside any argument that such division is indeed 'fair' as a philosophical construct, such division may adversely affect the in-business heir. Of course, if no heir is working in the family farm business, an equal division may indeed be fair.

As one Iowa farmer stated, 'No farmer ever bought an acre of land as an investment. We buy it as an income producing asset for the farm business. The non-farming heirs think their parents bought it as an investment.' What he meant by this is that the increase in land values, in some cases 20 to 30 times its value when the parents came into possession of it, causes the non-farming heirs to want their 'fair share'. While the old saw that, 'Blood is thicker than water' may still be true, it is equally true that 'Blood is not thicker than money.'

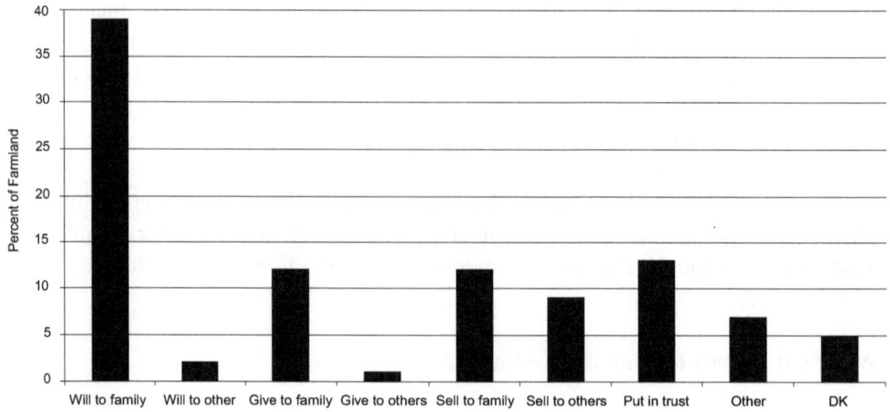

Figure 8.1 Anticipated method for transferring Iowa farmland

Source: *FARMTRANSFERS*

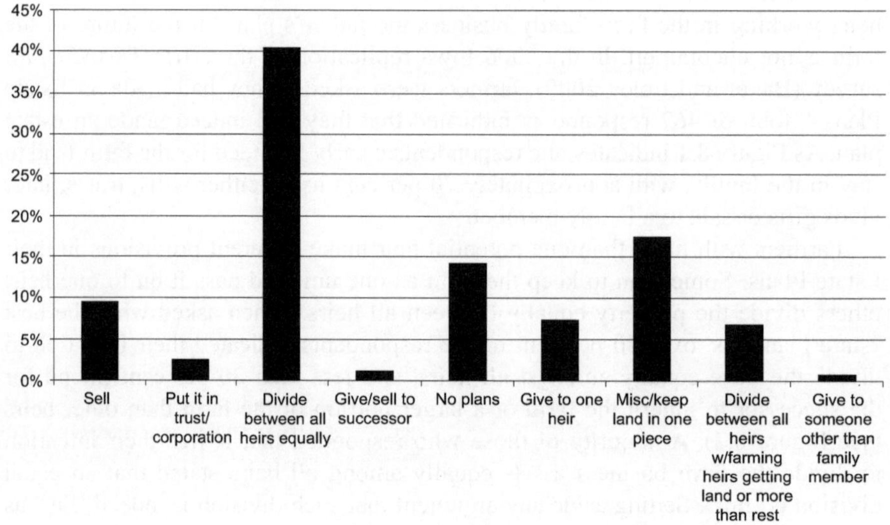

Figure 8.2 Farmers perception of the best plan for estate planning

Source: *FARMTRANSFERS*

When the asset value of farms and farm land experienced low appreciation the division of assets equally did not present a problem in designing an estate plan that was consistent with passing the farm family business to a successor. The relatively stable asset values and the high profit margins in farming allowed the successor to purchase the assets from the non-farming heirs without burdening the farm business with unmanageable debt. The value of farm assets, primarily land, remained stable from the end of WWII until the decade of the 1970's when land values began to rise rapidly (Duffy 2011). Concurrent with the rise in the value of farm business assets, primarily land, the profit margins in farming began to decline.

Table 8.5 Changing agricultural land values and net farm income in Iowa

Year	Weighted average value for all grades of Iowa farmland	Net farm Income per cent
1950	$218	93.3
1960	$261	60.3
1970	$419	59.1
1980	$2006	33.8

Source: ERS 2011

However, when there is an in-business heir, an equal division fails to recognize their contribution to the preservation or creation of wealth stored in the business. The heir may have either invested in the business or received modest compensation so that the business could grow and prosper. Unfortunately, such a contribution is rarely documented and a value placed upon it. Upon the death of the owner, the contribution of the in-business heir is included in the decedent's estate and divided among all heirs.

A second problem is that the in-business heir is forced into business with individuals not of his or her choosing and whose knowledge of modern farming may be woefully out of date. The non-farming heirs may not have the long-term vision of continuation of the farm family business, but may want to liquidate the farm business asset. Thus, as noted by the English farmer quoted previously, the farm business heir may need to work for many years to buy the assets from the non-business heirs. Such purchases may adversely affect the ability of the farm to acquire new equipment, technology or add new enterprises.

As the age of the owner–operators increases, the opportunity to buy the farm assets from the non-farming heirs may occur when the in-business heir is approaching retirement age, although they may be averse to taking on the debt necessary to purchase the assets. In such cases the farm land is often rented to neighbour and the farm business ceases to exist. With the passing of each generation, the heirs become more remote, that is the second and third generation heirs, and their portion of the inherited land may be so small as to be of little economic value.

The contribution to the wealth of the business is made in several forms. One is the *succession effect*, where the owner–operator generation decides to have a successor and begins the process of increasing the income of the farm to support a second generation. That process may, and usually does, include the purchasing additional assets or adding a new enterprise. Either of these will increase the wealth of the business.

A second form of increase in wealth can be attributed to the *successor effect*, where the successor joins the farm business and there is an excess of labour that must be employed. In this case, the business may add assets or develop new enterprises. In either case, the wealth of the business is increased.

A third form of contribution to the wealth of the business is the wealth that is preserved by the successor providing services to the owner generation as that generation ages. In many families, the successor may manage the farm business in lieu of the parents hiring a farm management company. The successor may also provide other services such as meal preparation, house cleaning, transportation for medical appointments, maintenance on the parent's home and other miscellaneous services. If the parents were required to pay for the services the wealth retained in the business would be decreased.

For those farmers who have a successor and are deciding how to divide assets in their estate plans, there are three basic approaches. One is to do nothing and to divide the assets equally among all heirs. A second approach is to 'freeze' the asset value at the time the successor enters the farm business. Usually, the value of the farm business assets is determined when the successor enters the business and any increase in the value of the assets after that point in time is credited to the successor. The justification for the increase in the appreciated value being credited to the successor is that, if there were no successor the assets could have been sold and the proceeds divided equally among the heirs. The third approach is to compensate the in-business heir for the wealth they have created in the business.

The following is an example of the third approach developed by Goeller (2008). Let us assume that in 1990 one heir returns to the farm business and the value of the farm was $600,000. Let us further assume that the two heirs are not working in the business. If the farm were sold and the assets were divided amongst the heirs on the basis of the contribution to the wealth held in the business each would receive $200,000 as no heir had added to such wealth.

Through growth, appreciation, inflation and diversification then assume that in 2010 the net worth of the farm business has increased to $1,500,000. The growth since 1990 is $900,000. The owners of the farm business estimate that one half of the growth is attributable to the in-business heir and believe that this contribution should be compensated. Therefore, the owners' estate plan divides the farm business in the following manner: the in-business heir receives one half of the growth, $450,000 plus one third of the of the owners growth, $150,000 plus one third of the 1990 value of the business, $200,000 for a total of $800,000. The non-business heirs receive one third of the owner's share of the growth, $150,000 plus one third of the 1990 value of the business, $200,000 for a total of $350,000.

Even when the owners provide for compensating the on-farm heir for their contribution to the owners wealth, it may be difficult for that heir to cash flow the purchase of non-farm heirs' interests. The owners may establish buy–sell agreements between and among the heirs. A buy–sell agreement is a contract obliging one business owner to buy all or a portion of the business upon occurrence of a specified event. In an estate plan the specified event is the death of the owner. The buy–sell agreement may set the price, length of contract, interest rate and other terms and conditions of the purchase of the non-farm heirs' inheritance. A buy–sell agreement prevents an unqualified off-farm heir from obtaining an ownership interest or from sharing in the management of the farm business. The agreement also avoids or reduces business disruptions when the owner dies because the event has been planned for and management will continue. Planning for the future in this way assures farm family business stability and continuity.

Creating a Farm Business Succession Plan

One of the most common questions received by the Beginning Farmer Center is, 'How do we get started?' One might be tempted to reply with the flippant answer, 'Start at the beginning.' Of course it is true with any planning process there must be a beginning and an ending. But for many farm families it is difficult to identify the starting point. This is especially true if the family has received a variety of advice which, in whole or in part, is conflicting. Even when farm families have attended seminars on farm succession planning and are prepared with accurate information, they may not be able to translate that information into the act of planning.

One effective method is to use the critical path method of management to develop a farm family business succession plan. The critical path method is a planning process that most farmers can adopt as it is an intuitive process. For example, if a famer is asked to list the steps in planting a crop they have little trouble in doing so. They may begin by having their financial records up to date and in good order, before they approach a lender for a crop input loan. They will order seed, schedule its delivery, order fertilizer and chemicals, perform maintenance on equipment and then begin field work. Farmers are familiar with the sequence of activities because that have planted crops dozens of times and they are familiar with the resources, scheduling of activities and priorities necessary to complete process of planning a crop. However, many, if not most, farm families have never engaged in the process of planning a farm business succession and are apprehensive about how to begin the process.

The critical path method, when used to develop a farm family business succession plan, requires the farm family to identify the necessary resources and identify and prioritize the activities necessary to complete the plan. It also requires an estimation of the amount of time for each activity, the start and end date for the activities and the total time for completion of the plan. Because of the identification and priority scheduling of activities, the planning is focused on those that are essential to complete the plan.

Using the critical path method is simple and straightforward. Farm families begin by listing all the activities that are needed to complete the succession plan. When the list is completed, it is arranged in the sequential order, beginning with the first activity and ending with the last activity. The first activity must be completed or near completion before the second activity can begin and the remaining activities must be completed in an ordered sequence (see Figure 8.3). Of course there are important activities that must be completed but are not in the sequential path. It is the context of the activity that will determine whether or not it must be included in the sequence.

It is difficult to accurately estimate the time necessary for a new activity and it is not unusual to underestimate the time needed to complete an activity. By using this method, if the time estimates are inaccurate, the plan can be adjusted and all involved in the planning are aware of the delay.

Once the list is complete, the activities are prioritized and ordered, with the earliest practical start date for an activity; an estimate of length of time to completion; if the activity is parallel or sequential; how the activity is to be measured; who is responsible for the activity; and how and to whom the activity will be reported.

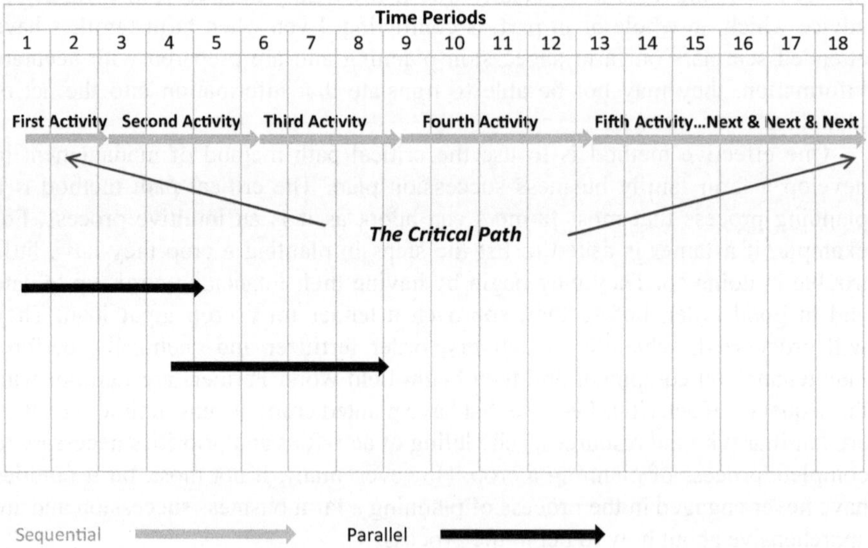

Figure 8.3 The critical path method

Figure 8.3 is an exemplar of a critical path chart. It is simply constructed by adding up the time of the sequential activities and captioning the top of the chart with the appropriate time periods, that is days, weeks, months or years. Each activity is indicated by an arrow. The position of the arrow indicates its priority in the sequence and the length is representative of the time needed to complete the activity. The critical path is the longest sequence of sequential activities leading to the completion of the plan. Of course, any delay in the commencement or completion of an activity on the critical path will delay the completion of the whole plan and the time allowed for future sequential activities will need to be shortened. Care should be taken to avoid scheduling conflicts between sequential and parallel activities.

Conclusions

The old saw that 'we don't know what we don't know' is surely applicable to farm business succession planning. A combination of experience in advising farming families and the *FARMTRANSFERS* research have highlighted both the complexity of farm business succession planning and the frequent lack of such a plan! The failure to effectively address these complex issues can and often does result in the failure to transfer the farm business to the successor.

The *FARMTRANSFERS* research has revealed evidence of a failure to understand, appreciate and address the key issue – the transfer of farm business assets upon the death of the owner. In the 2006 Iowa replication a majority of those surveyed responded that they intend to leave the farm to family members, and a majority of those who responded stated that the best estate plan is to leave the farm in undivided interests to all heirs. The owners in most cases had not considered the contribution of the in-business heir (who may feel aggrieved, if their siblings are awarded a share of the results of their labours) by adjustments based upon such a contribution.

As a result of this, the Beginning Farmer Center has developed a case study example, with a series of spreadsheets to assist in calculating the value of the in-business heir's contribution. Once the value has been determined and agreed, the estate plan can be adjusted accordingly to provide for an equitable distribution of farm business assets. This educational material and the mode and method of delivering it to farm families is a direct result of the research findings.

So what? Perhaps the ultimate answer to the question lies in the satisfaction and contentment of thousands of farm families who have developed and implemented effective and manageable farm business succession plans in recent years.

References

Baker, J.R. and Epley, E. 2009. *Iowa Farmers Business and Transfer Plans* [online]. Available at: http://www.extension.iastate.edu/bfc/pubs/IA%20Farm%20Business%20survey%20results.pdf.

Duffy, M. 2011. *2010 Farmland Value Survey Iowa State University* [online]. Available at: http://www.extension.iastate.edu/agdm/wholefarm/html/c2-70.html

ERS *Farm Income Data Files* [online]. Available at: http://www.ers.usda.gov/data/farmincome/finfidmu.htm

Goeller, D. 2008. *Putting a Value on Sweat Equity* [online]. Available at: http://digitalcommons.unl.edu/agecon_cornhusker/396/

Tannen, D. 1992. *You Just Don't Understand: Women and Men in Conversation.* Harper Paperbacks.

United States Census of Agriculture 2007 [online]. Available at: http://www.agcensus.usda.gov/Publications/2007/Full_Report/Volume_1,_Chapter_2_US_State_Level/st99_2_001_001.pdf

United State Census Bureau 2010 [online]. Available at: http://2010.census.gov/2010census/data/

University of Iowa School of Social Work 2009. *National Resource Center for Family Centered Practice Committed to Excellence Through Supervision* [online]. Available at: http://www.uiowa.edu/~nrcfcp/training/documents/Participant%20Packet%20Intergen%20Dynamics.pdf

Chapter 9
Facilitating Succession and Retirement in US Agriculture: The Case of Nebraska

Dave Goeller

Introduction

The University of Nebraska began educating farmers and ranchers in the area of farm business succession planning over 25 years ago. During that period over 1,000 Nebraska farming businesses have received assistance transferring their farm business from one generation to the next. Early efforts focused on the farm financial aspects of the transition process, however, because of farmer evaluations, educational efforts have shifted focus to communication skills and the importance of the business succession planning process.

Trends in United States agriculture over the past 25 years indicate the importance of farm business succession planning and the urgency created by increased concentration of land ownership in fewer and older hands. According to the US Agricultural Census taken by the National Agricultural Statistics Service, the average age of US farm operators has increased from 50.5 in 1982 to 57.1 in 2007 (United States Department of Agriculture 2009). While the number of total farm operators has only decreased by 36,184 or 1.6 per cent, operators over the age of 65 have increased by 256,058 or 64 per cent. Therefore, the over 65 age group as a percentage of all operators has increased from 17.8 per cent in 1982 to 29.7 per cent in 2007 (United States Department of Agriculture 2009). During this same time period there were 237,533 or 66.7 per cent fewer farm operators under the age of 35, which, as a percentage of all operators, indicates operators who are under 35 years of age have decreased from 15.9 per cent in 1982 to 5.4 per cent in 2007 (United States Department of Agriculture 2009).

Dr Michael Duffy, Director of the Beginning Farmer Center at Iowa State University, has researched the concentration of Iowa land ownership by those 65 years of age and older. His data, published in "Farm Succession in Iowa", 2000 indicates land owned by those 65 and older has steadily increased from 12 per cent in 1920 to over 55 per cent 2007. Similar trends have occurred throughout much of the Midwest United States.

There are several barriers preventing younger people from entering production agriculture in the US including the high capital costs of land, machinery and other farm assets, the increased mechanization of farming, US tax policy, tight profit margins for many sectors of agriculture, ease of operation of modern farm equipment, pride of ownership, love of the land, the "no one can do it as well as I can" attitude of current farm business owners, deficiency in communication skills, lack of retirement planning by the current farm operators, and lack of planning for a successor (Ahearn and Newton 2009). Although many barriers exist, interest from younger "want to be" farmers remains high. Many US states have linking services that match farm owners with prospective successors, e.g. *International Farm Transaction Network: Fostering the next generation of farmers* at: <http://www.farmtransition.org/> and the *Center for Rural Affairs Land Link program*. at: <http://www.cfra.org/resources/beginning_farmer/land_link>, and all linking programs indicate the number of younger potential farmers is several times larger than the number of older farm owners willing to participate in the linking programs.

A farm business is more than only the farm assets. The farm business includes people, the assets and liabilities of the farm business finances, the day-to-day operation, the marketing and purchasing as well as the labour, management and decision making aspects. Owners of a farm business make either a conscious or an unconscious decision regarding whether or not to actively pursue a successor for their business. Many farmers do not consciously consider the question, "Do I want my farming business to continue after I am gone?" They simply go on day after day doing their work as best they know how. One day simply turns into the next which turns into a week, and then a month and eventually 50 years have gone by and the question is asked, "Where has all the time gone?" Without thoughtful planning for succession, the unintended consequence is that there are fewer and fewer young farmers entering the business and ownership of land becomes concentrated into fewer and older hands.

If a generational transfer of a farm business is to take place it usually takes effort and planning. Bringing a successor into any business is a process not an event. It requires time, thoughtful discussion, and planning. Transition typically does not happen automatically. To begin the planning process, considering the following questions will help to establish some topics for discussion.

Questions for the owner generation:

- Do you want your farming business to continue beyond your life?
- Is it important to keep the farm in the family?
- Do you want to have a successor?
  - Have you identified a successor?
  - Is the successor one of your heirs?
- Will you retire?
  - If you retire where will you live?
  - How will you spend your time?
  - What kind of lifestyle do you expect in retirement?
  - How much money will you need to support that lifestyle?
  - What will be the source of your income?
  - Rental of farm assets?
  - Non-farm investments?

  - Sale of farm assets?
  - When will you retire?
- What are your expectations regarding:
  - The younger generation's income?
  - The younger generation's vacation?
  - The beginning and ending of the work day?
- What assets and liabilities make up the farm business?
- What is the profit history of your farm business?

Questions for the successor generation:

- What are your expectations regarding:
  - Your income?
  - Your vacations?
  - Your lifestyle?
  - Your health insurance?
  - The beginning and ending of the work day?
- What resources do you bring to the business?
  - Skills?
  - Experience?
  - Assets?
  - Desire?
  - Work ethic?
- Which of your skills needs improvement?
  - Analytical expertise?
  - Farm financial management?
  - Risk management?
  - Production/enterprise skills?
  - Communication skills?

The generational transfer plan of a farm business can be thought of as a three legged stool. The three key elements need to be interdependent so they complement each other rather than conflict. The three legs are a retirement plan, an estate plan and a business succession plan. Each leg is needed to support the successful farm business transition from one generation to the next.

The Retirement Plan

Lack of good retirement planning has become a major barrier to farm business succession. For many small business owners, and especially for farmers, the business is the key source for retirement income when or if retirement occurs (Errington and Lobley 2002). The land in essence becomes the pension plan with farmers using the land as an investment portfolio to be sold upon retirement.

Further, the reliance on the farm assets creates an uncertainty regarding the adequacy of retirement income. No one knows the answer to the question, how long will we live? It is very tempting for farmers to just keep on farming. Most enjoy their work and, with modern farm equipment, it has become very comfortable for most farmers to simply farm for one more year. Certainly no one should tell farmers they must retire. It is their business. They have worked hard for it, and they should do as they see fit. There are consequences, though, if a farmer continues farming until death; the likelihood of a younger operator stepping in and continuing the farm business is not high. Odds are at the death or disability of the older farmer, if a successor has not been identified, the family will either rent or sell the property to the highest bidder. Most of the time the highest bidder will be a large established farmer that will likely absorb this farm into their own. The unintended consequence of not planning to retire is there will be one less family living in the community, one less family attending churches and sending children to the local schools, and one less family buying goods and services from the local business. The unintended consequence is a less vital and vibrant community. A good retirement plan should address the approximate date for retirement, sources of retirement income, and retirement activities.

The Estate Plan

The first question owners of agricultural assets need to consider when doing estate planning is: "Do you want your farming business to continue on beyond your lifetime?" If the answer is no, then simply decide who you want to leave your assets to, when you want to leave it to them, and how long you would like to exert control over those assets. Go to your attorney and draw up your estate plan. If on the other hand the answer to the question is, "Yes, I do want my business to continue on," then you must consider who will be the successor to the business? If the successor is an heir, will that heir be treated differently than the other heirs? Will the farm assets be passed down in a financially viable unit or will they be cut into pieces and divided up like a pie?

We will all die eventually, and, in the case of farm owners, at the time of death their farm assets will pass down to their heirs. It is possible for an heir to take over an operation without previous involvement and still become successful, but one can greatly increase the probability of success if the owner plans for a successor.

One of the most difficult decisions for farm families to make is the one regarding asset distribution in their estate plan. Most of us love all our children, but we know that as they were growing up, they were not treated exactly the same. If one child needed to have an appendix operation we didn't rush all the kids to the doctor to keep things even. Each was probably given what they needed, when they needed it, to the best of our abilities. However, many of us have a preconceived notion that if there are five children in the family the estate plan should divide all assets five equal ways.

The following is a recap of an actual conversation with a 68 year old farmer in Nebraska named Joe.

> After a rather lengthy conversation with an older farmer discussing the importance of the successor generation gaining experience in management and decision making, the 68 year old farmer named Joe asked this question, "You're the expert; at what age do you think a father should start sharing management duties with his son?" "Well, Joe, how old is your son?" I asked. "Ah, my son is 39 years old, but I wasn't asking about my son, I was asking about my father."
>
> The father/grandfather in this family was 90 years old, and he had maintained complete control of all management and decision making authority. Joe then spoke about their typical day. He said, "Each morning my son and I drive over to my dad's house and wait in the farm yard in our pickup truck until Dad has finished his breakfast. He then leaves the house and walks over to us and explains what we are to do for the day. For example he looks at me and says, "OK you go feed the cows." and to my son, "and you go rake hay." He tells us what to do and when to do it."
>
> What are the chances that someday when the 90 year old father/grandfather dies that Joe or his 39 year old son will have the management skills required to profitably operate a modern farm business?
>
> After some further discussion, I asked "What is your father's estate plan?" Joe replied, "Well I really can't say. We have never discussed it. We don't talk about stuff like that much." A few months later on a return visit, the topic of the estate plan came up. He said, "Dave, I have some bad news. I asked my Dad about his estate plan, and he told me that he plans to divide all of his assets equally between my seven brothers and sisters and myself. So I will get 1/8 of the farm just like each of my siblings who have not spent a single day working on this place since they left home." He had worked on that farm for over 50 years. He was paid a very low wage and given a sub-standard house to live in plus a half a beef per year.

This true story does not have a very happy ending. The 39 year old son, who was also being paid a very meagre wage and a half a beef per year, left the operation shortly after discovering the facts of the estate plan. He has acquired a good job and is doing quite well, however, Joe did inherit only 1/8 of the farm at the death of his father. He tried to acquire financing to buy out his brothers and sisters but was unable to do so. The farm was sold to the highest bidder who happened to be a large landowner in the county. Joe and his wife have unfortunately had to spend most of their inheritance on hospital and doctor bills because of health issues. They currently reside in a low income public housing project in a small Nebraska town and are living on their social security benefits.

Was Joe treated fairly? He was certainly treated equally. It has been said, "The most unfair thing you can do is to treat un-equals, equally." Shouldn't the compensation be comparable to the contribution? Should farming heirs who helped build the farming business, contributing 50 years of their lives living on a sub-standard wage, be treated equally with their seven siblings, who only show up for Christmas and the funeral, be treated equally? Is equal fair?

This true account of one family's situation illustrates the need to develop some way to value the farming heir's contribution to the business. This concept is "Contribution Should Equal Compensation".

Let's take a look at another family. Let's call this family the Jones. The Jones' have three children. The youngest of the three children is Jimmy. He has decided to come back to the farm. The other two siblings chose not to be involved in the family business and have chosen professions outside of agriculture and moved to neighbouring states. Jimmy came back into the farming business in 1990. At the time of his return mother and father Jones valued their estate at $300,000. They felt all children had contributed equally throughout their growing years so at the time of Jimmy's return to the farming business, Dad and Mom feel equal would have been fair. If they had passed away in 1990 each child would have received $100,000. Twenty some years later much has changed. Jimmy has been compensated with a somewhat low wage, but, because his wife works away from the farm, they have been able to get by. His contributions have been significant. Jimmy encouraged Dad to buy the neighbouring farm across the road when it came up for sale. If Jimmy had not come back to farm, Dad would not have been able to handle the additional work. Because of Jimmy, Dad purchased the additional farm. It has appreciated in value significantly since that time. Jimmy was also responsible for introducing a beef cow enterprise to the farm and has been largely responsible for the genetic selection that has led to improved profits for the beef herd. Jimmy has taken over much of the management and decision making the past several years. Dad and Mom still receive significant income from the farming business and will need to continue to do so as they have not developed other retirement strategies.

Today Mr and Mrs Jones are considering how to structure their estate plan. Their estate has grown from $300,000 to $3,300,000. The business growth since Jimmy's return is $3,000,000. The question they are considering is, "How much of that appreciation, reinvestment of profits, growth and success has been due to the fact that Jimmy came back to the farm in 1990?" Mrs Jones reminds her husband that they probably would not have purchased the additional land had Jimmy not returned to the business. In fact, they may have needed to sell some of the original land to help support income needs in their older years. Jimmy has also had a huge impact on the beef enterprise. It has contributed significantly to the overall profitability of the business. Without Jimmy's input in both labour and, more recently, management of the farming business, the current size of Mr and Mrs Jones' estate would undoubtedly be much less. After careful thought and consideration Dad and Mom decided that Jimmy was responsible for 50 per cent of the business growth, and Dad and Mom were responsible for 50 per cent.

As the Jones calculate how to make a plan that accounts for Jimmy's contribution they first consider the 1/3 of the asset value of their estate when Jimmy came into the farming business; 1/3 of $300,000 is $100,000. They then consider Jimmy's 50 per cent contribution to the growth, appreciation, and profit reinvestment of the business since his return; 50 per cent of $3,000,000 is $1,500,000. Finally, they consider 1/3 of the 50 per cent contributed to the business by Dad and Mom; 1/3 of $1,500,000 equals $500,000. Thus, Jimmy will receive an inheritance of $100,000 + $1,500,000 + $500,000 = $2,100,000, while his siblings will receive $100,000 + $500,000 = $600,000. This is certainly not equal, but, based on their situation, the Jones feel it is fair. Contribution equals Compensation.

A farming estate plan must also account for the fact that the returned earnings of US farm assets are not traditionally very attractive to non-farming heirs. Because asset values, especially farm land, have increased rather dramatically in the last 20 years, the value of many farm estates is relatively high. The earnings those farm assets can produce, however, is modest. The problem arises when non-farming heirs try to stay in business with farming heirs. The income return non-farming heirs will receive on their farming inheritance will not be comparable to many non-farm investments without capturing asset appreciation through the sale of the farm assets. If possible, it may be better to provide an inheritance for non-farming heirs from assets that are not crucial to the continued success of the farm business. Leaving non-farming heirs non-farm assets such as life insurance, houses, stocks, bonds, or other non-farm investments will not jeopardize the continuation of the farming business.

The Business Succession Plan

Good communication skills are vital to creating and executing a viable farm business succession plan. Transitioning a farm business from one generation to another involves much more than simply transferring ownership of the farm assets. It involves passing on the skills and knowledge the owner has acquired over time and the ability to manage the farm business profitably. A crucial skill that many farmers lack is the skill of communication. The process of developing a successor requires many hours of discussion between the generations. It should include all actively involved parties, especially spouses. A key element that needs to be established is the expectations of each generation. If expectations are not discussed and revealed, future problems are very likely to erupt. A simple, useful exercise is to ask each individual to write a sentence or two briefly describing his/her personal response to the 12 questions regarding his/her individual expectations concerning the farm business succession process.

Expectations to consider:

1. What time does the work day begin and end?
2. What about noon hour?
3. Will there be a vacation?
4. What about weekends?
5. What are the expectations regarding life style?
6. How much income is needed by the successors? Family living expenses vary tremendously from family to family. If a successor has the expectation that they will need $60,000/year to provide family living expenses, where will that come from? The owners also have expectations regarding family living. If they are expecting that a portion of their living expenses will come from operations of the farm business, will the income be adequate to provide two family livings?
7. How long will the transition period last?
8. Does the older generation plan to retire?
9. If so, what is your vision of retirement?
10. Will the owners eventually move from the farm and, if so, where to?
11. Do the owners desire to continue involvement in the business after retirement?
12. If so, how will that look?

All parties to the farm business succession, including spouses, should discuss their expectations. Where expectations differ, the ensuing discussion can provide an opportunity to develop joint problem solving and communication skills that will be essential to an effective generational farm business transition. If disagreements can be resolved civilly to the satisfaction of both parties, the development of agreements can begin.

A proven business succession planning model is available at <http://www.extension.iastate.edu/agdm/wholefarm/html/c4-10.html>. This was developed over the past two decades by Iowa State University and the University of Nebraska, and it considers the typical farm business to be composed of labour, income, management, and asset ownership. Each of these business components will need to be transferred from one generation to the next. Because the transferring of business components may be difficult to reverse, it is suggested transferring occur in phases. This model involves four phases of transition: the testing phase, the commitment phase, the established phase, and the withdrawal phase.

The Testing Phase

The testing phase can be thought of much like dating. It is a trial period designed to explore compatibility of the parties and to determine if going forward with a business partner relationship is desirable. It is particularly important, if the successor is also the owner's child, for the owner to make a conscious effort to develop a business partner attitude and relationship with the successor as opposed to a parent/child relationship. Consider the parents that are teaching their child to walk. Dad and Mom each take hold of one of the toddler's hands and gently encourage the first step. As the child matures, if Dad and Mom continue to hold a hand as the child grows, he or she will likely never fall, however, neither will the child learn to walk or to run independently. Much in the same manner, a parent needs to consider developing a successor that will continue the legacy they have developed. The owner attitude shifts from one in authority, to one as a mentor, coach or teacher, instructing and guiding, but allowing for mistakes and falls.

In the testing phase a detailed outline of labour duties should be discussed. A clear job description should be agreed upon. During the testing phase it is often desirable to simply create a wage arrangement to provide income to the potential successor as opposed to making any long term arrangements that may be difficult to undo should the business succession process be terminated.

Management is probably the most difficult business component to transition from one generation to the next. Giving up managerial control of the farm business is a very difficult thing for most owners, therefore creating a management transfer plan is critical. Most farm business succession plans will not transfer any management during the testing phase. Creating opportunities to allow the successor some insight as to how management decisions are made will be very helpful as the process continues.

There also should not be any transfer of farm business assets during the testing phase. It can be very expensive to reverse asset transfer because of tax consequences and the potential for legal battles, if the succession plan is abandoned, is heightened.

The length of the testing phase should be relatively short. One to two years is typical. A predetermined meeting date should be chosen to mutually evaluate the possibility of moving to the next phase. If, after meeting, all parties agree they would like to continue, they then each make a commitment to proceed.

The Commitment Phase

During the commitment phase both generations make the decision that the testing phase has been completed, and they each choose to move forward. The younger generation makes the commitment that they plan to make this farm their way of life. The owner generation makes the commitment that they desire the business to continue and commits to the process.

The commitment phase will challenge communication skills for both generations. One method that has proven effective to enable communication is to have regular farm meetings. One of the first items that must be addressed early in the commitment phase is the type of joint farming arrangements to be pursued. All parties must identify which type of joint farming operation they want to create. There are two basic types of joint farming arrangements, a "spin off business" and a "super farm business".

Under a "spin off" type of arrangement, the size of the existing operation is not adequate to provide income for two families. The successors will begin to create their own business by renting additional land, developing a completely new enterprise or supplementing the farm income with a non-farm job. The owners of the original farming business will continue to operate their farm as they have in the past with the exception that they will share machinery and trade use of machinery to the successor for the contribution of labour from the younger party. Both the successor and owner can buy inputs together to take advantage of increased quantity discounts and will work together in other ways that make sense to both parties. This arrangement will continue until the older party decides it is time for retirement at which time the younger party will take over both operations.

The "super farm" type of arrangement would best be described as a larger operation that has adequate income to provide for both families. Many times a larger "super farm" type has had hired employees that now can be replaced by the successor. The successor will be compensated with a wage, at first, followed up by an incentive arrangement with the option to rent additional land if desired. When the owner retires the successor assumes the operation of the farming business.

It is important to discuss both the type of joint farming arrangement as well as the duration of each of the three remaining phases during the early commitment meeting. Creating a timeline for each phase will provide security to the successor as well as a target date for the owner.

Income during the commitment phase should be derived at least in part from the success of the farm business. There can be a wage plus an incentive based on profits, yields or bench marks, or the successor may choose to begin an enterprise on their own. That could be renting some additional land or developing a companion venture.

Labour duties should be reviewed and modified as needed in the commitment phases. A clear description of all expectations should be reviewed and agreed upon.

A detailed management plan describing how management decisions will be made and who will be responsible to make them is critical. A good example of an effective management transfer plan can be illustrated by the plan a Nebraska family developed. The Howard family has their son James returning to the farm. They have a rather large operation and James will be replacing an employee. As they discussed management and management duties, they decided the remaining length of their farm business succession plan would be 15 years.

- For the first five years, Dad will continue to make all the management decisions, but when he goes to the bank to discuss financing, Dad will take James along to observe and learn. When Dad goes to the seed and fertilizer dealer, James will be included. When Mr Howard markets their cattle, James will participate and make the needed business contacts so vital to the business. Dad will discuss with James how he makes the decisions and why he has chosen specific options. Dad will prepare James for success by sharing his years of valuable experience with his son.
- The second five year period Dad and James plan to make the management

decisions together. They plan regular meetings to review alternatives and discuss management strategies.

- The last five year period James will be the primary management decision maker. Dad will still be involved in the business and will be available if James requires his assistance, but James will take the lead.

Success of the farm business is critically reliant on the management decisions. There is no room to allow self-esteem or pride to interfere with development of a successful farm manager to continue the farm business operation.

Early in the commitment phase there needs to be an open and frank discussion regarding asset and liability values of the existing farm business as well as dialog addressing how and when asset ownership will be transferred. Both parties may choose to have ownership of farm assets begin to be transferred during the commitment phase. Discussion of farm business structures such as Limited Liability Companies, Corporations or Partnerships may be appropriate tools to consider at this juncture. Tax consequences as well as availability of income must be considered if asset transfer methods include sales or gifts. It is critically important to document and properly record all asset transfers. If new purchases for the farm business are to be made, it may be beneficial, if income will allow, to have the successor make the purchases as opposed to the older generation purchasing assets.

If the successor is a family member, this commitment phase may be a good time to inform other family members of the arrangements that have been created and commitments that have been made. Although there are no two families alike, surprises usually cause problems, especially when inheritance is involved. If the successor is a family member, make sure to record all agreements in writing as required by law. If the successor is not a family member, it is equally important to make an extra effort to insure all agreements are in writing and recorded as required.

Ownership of farm business assets may begin to transition to the younger generation during this phase. If ownership transfer is to occur, the types of assets that should be transferred first are the income producing assets, such as leases for rented land, operating inventory, breeding livestock or new purchases of machinery.

The Established Phase

At the beginning of the established phase, schedule a meeting to review the plans and timelines made during the commitment phase. There will need to be adjustments and changes as plans and reality are seldom identical twins. During the commitment phase the successor generation will have gained valuable experience and knowledge and should be well on their way to becoming the operator of a farm business.

During the established phase, income for the successor should come exclusively from the farm business. Wages and incentives can be replaced by farm business profits. There may be situations in which income did not equal projections and, in those situations, adjustments can be made.

At this time the successor will be providing a large share of the labour. In most situations the owner generation will still be providing some labour, however, it may not be at the same level as the successor. It may become necessary to consider adding additional employees to the farm business.

Management of the business is critical during all phases but it can be particularly tricky to pass the management from one generation to the next. Transition of management usually begins to occur during the established phase. It can be very difficult for owners as they watch their successors lead the business in a direction that may be different than they would have chosen. Typically, as people age they become more risk averse (Bucciol and Miniaci 2007). If the successor makes management decisions that have the potential to expose farm assets to an increased level of risk, care should be taken to minimize exposure of the owner's assets. Taking risk may earn rewards. Likewise if rewards have been earned they should be a reward to the party exposed to the risks. Certainly the successors will make mistakes and, likely, different mistakes than the owner would have made. Owners need to be supportive, patient and encouraging when problems arise. Second guessing and "I told you so" comments will produce an unconfident successor who will be insecure in their abilities. The skills required to become a successful modern farm business manager include analytical expertise, shrewdness, persistence, competitiveness, production knowledge and a relentless work ethic. Some or all of these skills have been effectively developed by many farm business owners over the years. These same skills, however, can be very detrimental as the owner begins to relinquish management authority to the next generation. Throughout this phase, it will be important for the owner generation to keep their final goal in mind, their farming operation continuing successfully.

During the established phase, transfer of asset ownership should be underway. There are numerous tax consequences asset transfer can trigger so tax consequences must be carefully considered. Basic asset transfer methods include outright sales, instalment sales, gifting and time of death transfers using such tools as wills and trusts. As farm business assets are transferred from the owner generation to the successor, it is critical that the transfers are consistent with the owner's estate plan.

The Withdrawal Stage

The withdrawal stage is the final step in the process. At the beginning of the withdrawal phase, schedule a meeting to review the plans and timelines made during the established phase. Make needed adjustments to the farm business succession plan where needed. The successor should now have both the experience and the skill needed to be in position to take over the farm business.

Income planning will be critical for the owner generation at this point. An accurate estimate of family living expenses in retirement must be developed to ensure that income will enable the lifestyle desired by the owners. If non-farm retirement income sources combined with farm rental income are not adequate to provide the desired retirement income it may be necessary to sell some of the farming assets to the successor. Cash flow for the successor is also critical. This segment of the generational transition process can prove very challenging financially for those farming businesses that have not provided for any non-farm retirement income. It may require adjustments to both the pace of farm asset transfers and the estate plan. If changes to the estate plan are required, consider the contribution of the purchaser and the purchase price when determining the level of compensation.

Labour during the withdrawal phase can also become a bit strained. A clear understanding of the type and amount of labour contributed by the owner should be discussed. One family in Nebraska, the Thompsons, held a family meeting discussing the contributions of the father Larry to the farming business. Larry stated, "I still want to contribute some labour to the farm business but retirement to me means that I will only do the work that I want to. So when I am driving the combine and it breaks down, I want to call my son Dean and tell him the combine broke. I am going to town to have coffee. Give me a call when you have it fixed." Retirement has different meanings for each individual. It is important for the older generation to consider their vision of retirement. What will retirement look like? Are there things that you have always wanted to do but never had the time? Where will you live? How about travel?

Management of the business should be completely transferred to the successor during the withdrawal stage. If difficult situations arise, the owner is still available to consult and give advice if desired.

Ownership of farm assets that are planned to pass during the lifetime of the farm owners should be well underway during the withdrawal stage. The present US capital gain tax consequences known as a "step up in basis" encourage owners of capital assets to postpone asset transfer until the time of death of the owners. Another consideration that further complicates farm asset ownership transition is that of long term care. If ownership of the farm assets remains with the older generation, they may be in jeopardy should an extended stay in a long term care facility be needed. Nursing homes can easily cost several thousand dollars per month. A prolonged stay can jeopardize farm assets and the overall economic viability of the farm business. A plan to address the risk management strategy regarding long term care should be considered. If long term care insurance is to be used, premiums become increasingly more expensive with age so it may be desirable to purchase long term care insurance early in the farm business succession process.

Conclusion

The US agricultural trend of greater numbers of older farmers owning increased amounts of farm land and assets, coupled with fewer younger farmers owning decreasing amounts of farm land and assets has occurred as a result of interaction between several factors. Many of those factors will not be changed. Although several states have legislated Beginning Farmer Tax Credit programs that encourage landlords to rent agricultural assets to beginners, the US Federal Tax Code discourages the sale of capital assets by the older generation.

Education has made and will continue to make a difference. Educational factors that can make an impact are:

- Teaching owners of farming businesses the importance of retirement planning and the unintended consequence of failure to plan for retirement.
- Informing farm business owners of the importance of planning for a farm business successor.
- Educating farmers regarding the farm business succession planning process.
- Educating farmers to create communication opportunities and encourage efforts to enhance communication skills.

Farm business succession planning is a process not an event. It takes time, effort, thought, calculations and communication. The process is critically dependent on each generation's ability to communicate. Educators have the opportunity to have a significant impact on farm businesses that are willing to participate in the farm business succession planning process.

References

Ahearn, M. and Newton, D. 2009. *Beginning Farmers and Ranchers*, EIB-53. Washington: USDA, Economic Research Service.

Baker, J. *International Farm Transition Network: Fostering the next generation of farmers* [online]. Available at: http://www.farmtransition.org/ [accessed 20 January 2011].

Baker, J. and Duffy, M. 2000. *Farm Succession in Iowa* [online]. Available at: http://www.extension.iastate.edu/bfc/pubs/FarmSuccession.pdf [accessed 20 January 2011].

Bucciol, A. and Miniaci, R. 2007. *Risk Aversion and Portfolio Choice* [online]. Available at: http://economia.unipv.it/eco-pol/PaperSeminari/Bucciol_Miniaci.pdf [accessed 20 January 2011].

Center for Rural Affairs [online]. Available at: http://www.cfra.org/resources/beginning_farmer/land_link [accessed 20 January 2011].

Center Programs: Beginning Farmer Center – Resources to Help our Next Generation of Farmers [online]. Available at: http://www.extension.iastate.edu/bfc/programs.html#Farm On [accessed 20 January 2011].

Errington, A. and Lobley, M. 2002. *Handing Over the Reins: A Comparative Study of Intergenerational Farm Transfers in England, France, Canada and the USA*. Paper presented to the Agricultural Economics Society Annual Conference, Aberystwyth, NSW, 8–11 April 2002.

South Dakota Department of Agriculture: Farm Link Program [online]. Available at: http://sdda.sd.gov/AgDevelopment/FarmLink/default.aspx [accessed 20 January 2011].

United States Department of Agriculture 2009. *United States Department of Agriculture: The Census of Agriculture* [online]. Available at: http://www.agcensus.usda.gov/Publications/2007/Full_Report/index.asp [accessed 20 January 2011].

University of Nebraska-Lincoln Women in Agriculture: Returning to the Farm [online]. Available at: http://wia.unl.edu/ReturningtotheFarm [accessed 20 January 2011].

Chapter 10
Retired Farmer – An Elusive Concept

Joy Kirkpatrick

Introduction

Driving the rural highways and country roads of the US, you can't help but notice the abundance of deteriorating buildings, silos, and disappearing fences. The farmstead just around the bend is still kept up, the flower beds are tended and the grass around the long-empty machine shed, livestock barn, and chicken coop is still mowed within an inch of its life. The vines are creeping up the silo, slowly pulling the structure down, but it is a beautiful destruction, with the wind rippling through the vine's leaves and giving the illusion the silo is slowly shimmying in the evening breeze to music only it can hear. There's a sedan in the driveway, it travels to the grocery store, the clinic, and church on Sunday mornings and Wednesday evenings in the summer. The farm couple or, more often than not, the farm widow still lives in the farmhouse, living on the rental income of the farmland and a small Social Security check. In most years this provides for her needs, with her small savings acting as a cushion for unexpected expenses. She hopes to be able to live her life in the farmhouse, avoiding the nursing home, if she can possibly help it. The aversion to the nursing home is two-fold, nursing homes are for the sick who can no longer care for themselves, but the expense, she must avoid the expense because all she has left is the land and that, above all, must go to her children. She sees them on holidays and her birthday, otherwise living her life in solitude, which is what she thought she wanted, didn't really consider any other life – once you marry a farmer you are committed to him and the life farming provides. She follows the generation that lived with their children as they aged, but birthed the generation that expects more in their retirement years, but are not sure how to achieve it. Strauss and Howe (1992) would classify this woman in the G.I. or Silent generation, and her children would be the ubiquitous Baby boomers.

Baby boomer farmers may have drastically different expectations for their retirement years as compared to their G.I. or Silent generation farming parents' retirements. This may be due, in part, to longer life expectancy but also due to the influence of their peers in other careers, who may have pensions or traditional retirement investments on which to rely. But the boomers may not be so different from their parents in their connection to the land, their love of farming, and the fact their identities are so connected to farming they find it difficult to think of leaving it. Recognizing when to step away from the business can be difficult to discern. Reasons for delaying retirement are plenty. Lack of retirement planning and dependency on the farm for income into the retirement years can be two major reasons. However, lack of planning or little retirement income can be used as a shield to avoid facing the emotional issues, such as losing a sense of identity and the emotional ties connected to the farming business.

Retirement planning is not a daily chore or a skill that can be used on a regular basis in a farming operation. In comparison, daily farm tasks and production management decisions can provide more immediate gratification to managers. For these reasons, it is easy to delay and avoid retirement planning. But now, as boomers continue to wield their power on the nation's attention, first with rebellion, music, and concern for the environment, now their attention is focused on their unease for their retirement years. With the first wave of boomer farmers in or nearing retirement there may be enough of a critical mass of farmers interested in planning for it, albeit last minute planning.

Retirement planning is essential to developing a sustainable family farm. Retiring farmers must answer the questions: where to live, what to do, how to fund it, and put the answers against the backdrop of the farm business continuing for the entering generation. This chapter will discuss the research on farmers' retirement patterns, such as the age farmers are retiring or say they plan to retire and comparing these patterns across countries. It will also discuss how the US Census of Agriculture data may obscure true retirement patterns because of the USDA definition of a farm and the phenomena of retiring *to* farming in the US. It will present a summary of research about where retiring farmers plan to live, and their retirement income sources and how, in the context of retiring from farming and farm succession, this has an impact on true transition of management of the farm. The chapter provides insight on the emotional connection farmers have with their occupation and in their own words illustrates the complex and competing emotions that can derail not only retirement but the succession of the farm to the next generation. Finally, the chapter outlines the need to move farmers from discussing the tangible details of retirement planning and immediately dismissing the possibilities to discussing the lifestyle they might enjoy if they recognize and process the emotional reasons for their hesitation in planning for their retirement.

Increasing Age of Farmers and Retirement Plans

There is little data on the retirement patterns of farmers; however, the US Agriculture Census and the Agricultural Resource Management Survey (ARMS) can give some insight on patterns by analysing the average age of farmers over several years. Farmers are considerably older than the rest of the US labour force. Over 25 per cent of all farmers, and about half of all agricultural landlords are 65 or older, compared with only about 3 per cent of the overall labour force. Older farm operators and landowners operate over one-third of all farm assets and are staying on the farm longer than previous generations (Mishra, Durst and El-Ostra 2005). The 2007 US Census of Agriculture noted that the fastest growing group of farm operators is those 65 years and older, with a 22 per cent increase over the 2002 Ag Census data. While the average age of the New Zealand farmers is considerably younger than farmers in the US, trends indicate the average age of the New Zealand farmer is steadily increasing (Fairweather and Mulet-Marquis 2009). The increasing average age of farmers seems to indicate that farmers are delaying retirement, but the data does not provide insight into the reasons for the increasing average age, retirement decisions or attitudes, or the sources of income farmers are considering in their retirement years.

Since 1991, researchers have been using the *FARMTRANSFERS* questionnaire to ask farmers important questions about farm succession, retirement and asset transfer. As Chapter 1 explained, this survey has since been replicated in several countries and several states in the US. The surveys asked questions about retirement plans, whether the farmer planned to fully retire, semi-retire or never retire, sources of retirement income, and if they planned to retire, where they would live in their retirement. In a development of the questionnaire in 2006, Iowa researchers clarified the definitions of retirement categories provided to survey participants to more clearly delineate full retirement from semi-retirement and never retiring (Baker and Epley 2009):

- Retire: you will provide neither managerial control nor labour to the farm.
- Semi-retire: you will provide some managerial control and/or labour to the farm.
- Never retire: you will maintain full managerial control and provide some labour to the farm.

The survey question of full, semi or never retiring was not connected to sources of retirement income, which would have complicated the definition. Questions about retirement financial resources were asked separately from labour and management issues. Baker and Epley (2009) found more farmers describe their plans as never retiring than those with plans to fully retire. A similar Wisconsin survey, conducted in the four southwestern counties with 589 responses (23 per cent response rate), found that 73 per cent of respondents from the Brucellosis Ring Test (BRT) list plan to either never retire or to only semi-retire from farming (Kirkpatrick 2006).

Baker and Epley (2009) and Lobley, Baker and Whitehead (2010) compared the Iowa data to past studies in the US and foreign countries and found that Iowans are more likely to either retire or semi-retire than farmers in other states. However, compared to farmers in other countries surveyed, American farmers in general are more likely to never retire. Japan is the only country that departs from this trend; with farmers who indicate very similar retirement plans to US farmers (see Chapter 4).

Foskey (2002) describes Australian farmer retirement patterns with three terms: *retirement in farming*, with the operator providing management, labour or both to the operation which is similar to semi-retirement, *retirement from farming*, (full-retirement) or *retirement to farming*. Retirement to farming is a form of retirement described as a farm operator who enters into farming later in life after retiring from a full-time job, or, as the farm grows and becomes sufficient, or debt is reduced, the operator can afford to leave an off-farm occupation.

Efforts to study farmer retirement and succession trends are complicated by this sector of farmers who are retiring *to* farming. Even the census data used must be scrutinized to clearly understand the differences between these sectors of farmers in the United States. One factor in the United States is the definition of a farm for Agriculture Census purposes. A farm is defined as a business that sold or normally has potential to sell $1,000 of agricultural products during the year. With this low threshold US Ag Census definition in mind, many farmers are retiring *to* farming rather than from farming. This may be one factor in the increasing average age of farmers in the United States. Farmers who are retiring *to* farms may not be as dependent on farm income for their family living needs because of Social Security, pensions, or other retirement savings garnered from their previous occupation. Farms where the older generation retired *to* farming, the familial, symbolic tie to the farmland may not be instilled in a child to become the successor. The children of the retired *to* farming operator may have no desire to farm, yet the operator may feel the farmland is the children's inheritance right and leave the land to his heirs. Even if the retired *to* farming operator is willing to allow the land to move outside the family, the farm may not be productive or profitable enough to entice a successor from outside the family. This increases this type of farm's potential for being a last generation farm. It may also put the land at greater risk for development or becoming exclusively recreational land. Younger beginning farmers who seek farming as a primary occupation and as a primary income cannot compete with the retiring *to* farming phenomena, which may raise or sustain land prices to unaffordable levels for beginning farmers with little capital.

This sector of retiring *to* farmers raises its own set of issues and research questions, but statistics support the view that the traditional farmer sector's average age is increasing as well. Reasons for farmers to delay retirement and actively engage in commercial farming longer can be as varied as the number of crops and products grown. New production technologies to replace physical labour have allowed farmers to farm longer. Improved health and longer life are assumed to play a factor in delaying retirement. A Swedish longitudinal cohort study on farmers and rural non-farmers sought to substantiate the theory that delayed retirement in farmers is due to better health in the farming population than the general population. However, Thelin and Holmberg (2010) rejected their hypothesis that farmers were healthier than other employees, hence contributing to their delayed retirement. They concluded that both health and lifestyle factors have little effect on retirement age for farmers and that social, financial, or cultural factors have greater influence on farmer retirement.

Farmers in other countries than the US may be more willing to retire because they are eligible for retirement benefits and incentives for retiring early. Examples of farmer early retirement programs are found in several European countries. Greece, Ireland and France are three countries that have the highest rates of participation in the EU farmers' early retirement scheme (ERS), with Norway, Finland and Spain participating to a lesser degree. Early retirement schemes encourage farmers between the ages of 55 and 66 to retire and transfer land to a younger farmer, either a beginning farmer or a younger farmer wanting to expand his operation. Bika (2007) notes that early retirement schemes appear to have assisted a number of elderly farmers to retire with dignity. Bika continues his critique, stating that ERS's ability to hasten structural change, such as lowering the average age of farmers, was not realized. Bika (2007) continues that the ERS's structural effect would have happened anyway, just over a longer time period, and the ERS didn't increase retirement rates and did not facilitate farm transfers outside of family lines. Bika further concludes the early retirement program was most successful in continuing the family farm and in stabilizing the rural population rather than its intended objective of improving competiveness and structural change in agriculture. Similarly, Ryan (1995) contends that the early retirement schemes have had only marginal success in assisting resource transfer and explains low participation by several factors including low retirement benefits being offered, restrictive requirements instituted in some countries and the cultural resistance to leave farming. In Finland, farmers' early retirement programs began in 1974, with several temporary programs following in subsequent years. These early retirement benefits were farmer-specific and connected to the level of pension insurance purchased by the farmer while actively farming. Pietola, Väre and Lansink (2003) found that the likelihood of farmers exiting farming by using the early retirement program decreases as the farmer ages and the highest probability for voluntarily leaving farming is immediately after the farmer reaches the eligible age (55 years). Early retirement schemes may benefit those farmers who have larger, fairly profitable farms and who have already identified a successor to transition assets to the successor earlier than he would have without early retirement benefits. Taking advantage of the early retirement program at the earliest possible time may help retain the successor, especially in areas close to urban centres with more off-farm job opportunities (Pieotla, Väre and Lansink 2003, Bika 2007).

According to the 2006 Iowa survey (Baker and Epley 2009), the average age of retirement or semi-retirement for the respondents is 67 years old, compared to 66 years for respondents from the 2000 Iowa survey. This is comparable to the Australian study's findings for retirement age of 65.6 years. Australian, Japanese and United States farmers appear to plan to retire at an older age than Canadian, French or English farmers (Barclay, Foskey and Reeve 2005). US farmers responding to the survey may base their intended retirement age on when they would be eligible for full Social Security retirement benefits, rather than on the basis of providing less labour or management to the farming operation. The slight increase in the average planned retirement age of farmers in Iowa between 2000 and 2006 supports this hypothesis because the eligibility age for full retirement benefits is gradually increasing, depending on birth year. People born from 1943 to 1954 are eligible for full Social Security benefits at 66. Eligibility age increases two months for each year after 1954 (66 years and two months), until 1959 (66 years and ten months). Those born 1960 and later must reach 67 to be eligible for full benefits.

In the US, the Social Security structure provides a disincentive for retiring early, regardless of occupation. Social Security participants can begin to receive retirement benefits as early as 62 years old, but are penalized for taking Social Security benefits early. Benefits are reduced by approximately 30 per cent of the full benefit if they retire at 62 rather than their full retirement age. In addition to receiving a reduced payment, benefit income is withheld if early retirement participants earn more than the set income limits. When participants reach full retirement age the income limits are removed, but they will continue to receive reduced benefits.

Where Will Farm Retirees Live?

Most employees retiring from a career do not have to consider the decision of whether they will leave their home at the same time they retire from their jobs. Farm owners who plan to retire and/or transfer the farm to the next generation have to consider this long in advance of their retirement. Farms are one of the few businesses in which the family home and family memories are tied so closely together with the business. Operations with livestock enterprises are especially cognizant of the need for the primary livestock manager to be near the farmstead during farrowing, calving and lambing seasons. According to surveys in the US (Baker and Epley 2009, Kirkpatrick 2009) of those respondents who planned to retire, a majority of them (55 per cent in Iowa and 60 per cent in Wisconsin) do not plan to move from their current home. A majority of those indicating they would be moving plan to move less than ten miles from the home farm. A farm operator's decision to remain in the current home can reduce their retirement income needs, since a retirement home need not be purchased, rented, or built. However, it can drastically limit the next generation's ability to fully manage the farm, if they have to live even a short distance from the farm. The retiring generation must also consider their ability to truly relinquish control of the farm if they are living in the farmhouse, watching the daily activities and judging the successors' decisions. The family home is where the family milestones are celebrated; children are born and grow to adults. The desire to remain in the family home is natural, and this natural desire must be balanced with the needs of the business to not only continue but to thrive for the next generation. If retirement income is dependent on the farm continuing, leaving the home may be a small price to pay for the business to thrive and sustain multiple family living needs. If the retiring generation does plan to leave the farm home, the true costs of living off the farm must be calculated and factored into retirement income needs.

Retirement Income Sources

In the past, retirement income was commonly described as a three legged stool. One leg of the stool is Social Security benefits; another is employer defined pension benefits, and the third is personal savings. This three-legged stool is rapidly changing to a two legged stool for many, as employer defined pension benefit plans are no longer as prevalent for employees. Approximately half of the US workforce can expect a private pension through their employer. However, the trend is to transition away from private defined-benefit pensions to plans that depend more heavily on employee contributions. Social Security accounts for more than half of the total income for about 60 per cent of Social Security recipients. Social Security is the sole source of income for about 20 per cent of all recipients. In comparison, Social Security is on average only about 13 per cent of income for farmers who are receiving benefits (Mishra, Durst and El-Ostra 2005). This small percentage of income derived from Social Security may be because the farm operators are still receiving a significant amount of income from farm operations, but it may also be attributed to the amount of self-employment tax the farmer paid over his working life. Farmers, like many self-employed business owners, have always contended with the two-legged stool. Iowa farmers responding to the *FARMTRANSFERS* survey (Baker and Epley 2009) indicated several sources for retirement income. Social Security was the most common source identified (50 per cent of respondents). The other responses were income from the farm (41 per cent), income from a private retirement plan (37 per cent) and income from other investments (29 per cent). Sale of land, livestock, and other farm assets were among the lowest categories identified for Iowa respondents. In comparison, Australian farmers' most common retirement income will be the sale of farmland and other farm assets, reflecting a trend of farm land ownership as a retirement fund to be tapped when needed (Barclay, Foskey and Reeve 2005). Wisconsin survey respondents were asked what percentage of their income would come from various sources. Wisconsin respondents also indicated slightly more reliance on the sale of farm assets than Iowa respondents, with the three top sources expressed as a percentage of retirement income identified as: 1. Income from sale of farm assets (38 per cent); 2. Social Security (17 per cent); 3. Income from the farm (15 per cent) (Kirkpatrick 2009). Australian respondents top three income resources expressed as a percentage of retirement income were: 1. Income from the farm (40 per cent); 2. Sale of land (35 per cent) and 3. Other sources (30 per cent).

Government payments or farm programs can also be a source of retirement income. Hoppe (1996) suggests that the Conservation Reserve Program (CRP) could be considered a component of retirement income for farmers. Hoppe notes that approximately 18 per cent of retired operators had land in CRP in 1993, compared to only 11 per cent of all farmers. Farms with retired operators accounted for 26 per cent of the CRP farm enrolments and 28 per cent of the enrolled acreage.

Over 70 years ago, the process of retirement for the older generation began with the younger generation living with the older generation in the farmhouse.

The older generation would transfer the labour first and then the management decisions. Finally, through gifting or inheritance, the property would be transferred. This arrangement satisfied the need for the younger generation to begin their farming career, and provided the older generation a home and care in their older years. Many farms had three generations living under the same roof for several years. This arrangement had at least the possibility of matching the income potential of small family farms to the family living needs without the need for large expansions in operations, since multiple generations were living together. This form of retirement/succession cohabitation is still prevalent in Japan, with multiple generations sharing incomes and homes. The 2000 Japanese Census reported multiple generation families constitute approximately 80 per cent of the farms (Barclay, Foskey and Reeve 2005). However, in the United States and in many European countries, societal expectations and family needs have altered this arrangement almost to the point of extinction. The older generation is living longer and healthier lives and when they get to the point of needing care it may be more skilled care than the family can provide. The trend for the older generation to live independently from their adult children began in the early 1940s when the first Social Security payments were distributed. This additional income to the retirees provided a small but important financial resource allowing them to live separately from their children. By the 1960s, Social Security was so accepted that its effect on living patterns is rarely mentioned (Troll 1971). Financial independence is only one factor that influences whether the retiree will live with the successor generation on the farm. Age and health of the older generation are also factors, but financial independence is an important factor, with even a modest increase in income greatly increasing the odds of the older generation living alone (Chevan and Korson 1972).

The expectation for separate households remains the trend in farm families today. The cost of living for both the older and younger generations puts greater financial expectations on the farm business. Unfortunately, many farm operators continue to make investment decisions as if the old pattern of multiple generations living under the same roof is the norm. Farm operators continue to re-invest farm profits back into the farm, rather than invest in off-farm retirement plans, even though many have tax incentives attached to them. This systematic re-investment back into the operation means many family farms control considerable wealth. In 2001, US farm households had an average net worth of $545,869 compared to $395,500 for non-farm households (Mishra et al. 2002). This physical capital can be classified into liquid and illiquid assets. These relatively illiquid farm assets are further complicated because they are, in many cases, indivisible and can be a large percentage, if not all, of the family wealth (Mishra, El-Osta and Shaik 2010).

Categorizing these indivisible, illiquid farm assets as retirement assets makes the need for identifying a successor, preferably a family member, a priority for many farm operators. Retirement from farming is closely tied to decisions of farm succession and transfer of the business assets. Uchiyama, Lobley, Errington and Yanagimura (2008) found that farmers in England and Canada who had identified a successor preferred semi-retirement than those in the US or Japan. Having a chosen successor makes it easier for the farmer to reduce his level of involvement. Having identified a successor may also influence the continued capital investment the operator is willing to infuse into the farm business (Potter and Lobley 1992). This continued capital investment can make the operation vastly more attractive to the successor. It also further ties the retirement income of the older generation to a successful transition of management to the next generation. Conversely, if a related successor is not identified, and the operator places a high priority of keeping the farmland in the family, the operator is more likely to delay retirement indefinitely. In many cases, the operator must continue to farm to garner income from the assets. Eventually, as the farmer ages and his health diminishes, he may reduce the labour requirements by eliminating livestock but continuing the enterprises with more mechanization, such as grain crops or hay. The farmer may opt to let the livestock facilities deteriorate, rent out the cropland and continue living in the farmhouse in hopes the land will eventually transfer to his heirs at his death, in spite of the fact the heirs will never farm the land themselves (Potter and Lobley 1992). This process may severely impact the older generation's retirement income potential, considering their farm business investments may be their only retirement assets. The only way to realize their return on investment is to continue farming or sell the farm outside the family at a fair market value, either as a working farm, recreational land or for development. Relying on the farm assets for income and keeping ownership of these assets until death can pose risks to the inheritance plan. Many times, the female partner in the older generation outlives the male farming partner. In these cases, the ability of the female to continue the farming enterprise may be limited; therefore she may have only the farmland's rental income for her living expenses, which may impoverish her, or place her in the position of selling assets for income. If the older generation becomes reliant on skilled nursing care, and Social Security, pension, and farm income do not cover the costs of this care, the farmer may be required to sell non-residential business assets to pay for this care. If family members are unable or unwilling to purchase the assets or pay for skilled care from their own resources, the dual goals of using the farm for retirement and keeping the land in the family are lost.

Emotional Ties to Farming – what Farmers will and won't Miss about Farming

Retirement can mean many different things to people. It can mean the absolute end to a career, a completely recreational lifestyle, volunteerism, even transitioning to a new career or job. However, many view farming as not only a career or business, but as a way of life. Considering the family home is an integral part of the farm, the retiring generation does not often consider leaving their family home, even if they do want to retire from the daily responsibilities of farming. Research can provide patterns and trends, identify retirement income sources and suggest government policies to encourage farmers to retire early and to transfer the productive assets to a younger farmer. However, research findings will do little to persuade farmers to change their attitudes about farming as a lifestyle, retirement or inheritance of the farm.

When asked about what they would miss when they retire or semi-retire, the most common responses are connected with lifestyle, described in several different ways. Several Australian farmers expressed the loss of being in the outdoors (Barclay, Foskey and Reeve 2005), with one stating:

> I will miss the solitude of the place, the views, the company of the animals, the night sky and the privacy (Barclay, Foskey and Reeve 2005: 28).

Iowa, Wisconsin and Australian farmers all noted the loss of an active lifestyle, open spaces, and the independence that farming allowed them to experience (Baker and Epley 2009, Barclay, Foskey and Reeve 2005, Kirkpatrick 2009). Some farmers expressed what they would miss as they retire in a way that embodies the essence of farming, such as the Wisconsin farmer who said he would miss:

> Gratification of watching my crops grow, seeing my herd improve, in general, being the caretaker of the land (Kirkpatrick 2006).

Some farmers may be able to experience this same type of gratification once removed by watching a successor be good stewards or caretakers of the land; however, this type of fulfilment would be difficult to surrender after years of nurturing the land. Another Wisconsin farmer articulated his sense of what he would miss about farming in terms of lasting marks a farmer can make on the land and the resiliency of quality work on the farm by describing his love of, of all things, fencing:

> Fencing. When[I was]younger and was able to fence, I loved that because what else can a man do, that if you put all posts and wire (4 or 5 barbs) in new that you can forget about for about 30 years, unless a tree falls over it or something runs through it. I have a fence here that I put in new in 1965 and it's not in too bad a shape now. That will be 41 years this September. Cattle have been against it all these years (Kirkpatrick 2006).

Another element of loss mentioned in the Wisconsin study was the loss of control, with one farmer saying it would be hard to give up control "knowing someone else will have the say so of what was once all yours."

The Wisconsin survey collected many comments reflecting the respondents would miss nothing because they never plan to retire or cannot comprehend a life of retirement or semi-retirement. This is expressed by saying they will never stop working on the farm, or they will always help out on the farm for their son/children/heirs. One respondent said that he would miss "… breathing" because he'll be dead when he gives up farming, which is the embodiment of the "dying with your boots on" creed of many farmers worldwide. Extension educators at farm succession education programs sponsored by University of Wisconsin Cooperative Extension find similar attitudes of the older generation in their workshops. One teaching activity that can get this discussion on the table uses materials as simple as a blank piece of paper and coloured markers. The *Retirement Fantasy* activity asks participants to draw a picture of their retirement fantasy. There are no financial restrictions to the fantasy and they must not use words to depict their retirement fantasy, they must draw it – even if it is mostly stick figures and very abstract. When participants in these programs are asked to describe their "retirement fantasy", the older generation struggles to put even one retirement idea to paper, or if they do, it is a more relaxed version of what they are currently doing in their work life on the farm. In contrast, the successor generation, generally has little hesitation in describing elaborate retirement goals. Whether this difference is due to the naivety of the younger generation, resistance to change or inability to look past reality to envision a retirement fantasy on the part of the older generation, or a true generational shift in attitudes on retirement is yet to be determined in this anecdotal observation. However, the expectations of similar lifestyles to their non-farming classmates and peers is continuing to influence the next generations of farmers, including expectations of pay, work load, vacations and perhaps this extends to retirement expectations as well.

In the same surveys in which farmers are eloquently expressing their love of farming, they are still willing and quite capable of listing several things they would not miss about farming if they were to retire or semi-retire. The Iowa survey respondents' most common answers to the question of what they would NOT miss were: grain marketing, long hours, hard work and inclement weather (Baker and Epley 2009). Australian farmers expressed similar sentiments about the long hours, hard physical labour and working in extreme weather conditions (Barclay, Foskey and Reeve 2005). The Wisconsin farmers noted they would not miss the work load or schedule, with a large number of respondents reporting they would not miss daily milking and other daily chores (Kirkpatrick 2006). While not as common, respondents also noted financial uncertainty and risk were aspects of farming they would not miss.

University Cooperative Extension Services can develop educational programs to help farmers organize and implement their retirement and succession plans. These retirement education programs and materials, specifically targeting farmers, focus on the tangible decisions to make, such as estimating how much income will be needed in retirement and providing retirement savings strategies, resources, and vehicles. These resources are important and extremely helpful for retirement planning. But these programs are assuming the farm operator has made the conscious decision to retire or at least semi-retire from farming. Based on the *FARMTRANSFERS* surveys cited here and observations made by educators working with family farms on succession issues, this may be a faulty assumption. The concrete decisions that can be made in succession planning and retirement planning are easier to discuss. The inability to recognize, analyse and discuss the emotional aspects of retirement and succession can stall the process. Farm operations that would be considered financially sound, well-managed businesses can slowly collapse and fail because the older generation is unable or unwilling to face the contradicting desires of providing a legacy to his heirs yet retain the independence and lifestyle farming provides. Recognizing the long-term goals in terms of business succession, retirement decisions, income needs, and inheritance issues and analysing where these goals intersect and contradict can provide a platform for deeper communication, and understanding among the farming partners and off-farm heirs.

Conclusions

Farmers' decisions to never retire or only semi-retire and the increasing number of people retiring *to* farming are impacting the next generation's ability to embark on a true career path of full-time farming. While US Census data can be deceiving because of the low threshold definition of a farm, research data does support the proposition that the farming population is older than the work force in general. Low profit margins and the farm's inability to sustain multiple households are common reasons cited for delaying succession planning or ignoring the issue completely. Some studies show the lack of an identified successor is often a reason for delaying retirement. The timing of identifying a successor is critical for the business cycle of the farm. If the successor is identified, the older generation can be motivated to continue capital investments to assist the financial viability of the farm for the next generation. However, these continued investments into the farm business may make the older generation more reliant on and vulnerable to the next generation's ability to successfully manage the family farm. If a successor is not identified at the critical time, the older generation may slowly deplete the investments, and the farm gradually declines in value from the perspective of the next generation. The other component with timely identification of a successor is the infusion of Social Security income when the older generation reaches an age to receive benefits. The monthly income from Social Security and the addition of health care benefits through Medicare can provide just enough financial security to the older generation to be less reliant on a successful transition to the younger generation. Income from the Conservation Reserve Program (CRP) can have a similar affect, but goes one step further by taking land completely out of production that might have otherwise been rented to a beginning farmer or a farmer expanding his operation. The older farmer can eliminate the livestock, greatly reducing labour requirements and physical injury risks, and continue to live in the farmhouse, with no need to consider other living arrangements until his health greatly diminishes. These are all tangible aspects that can be quantified and analysed. Policies can be developed and programs piloted to mitigate risks to the older generation's financial stability. These may work to encourage a slightly earlier exit from farming, but may not be incentive enough to entice a significant percentage of farmers to completely retire. The value placed on lifestyle quality, the sense of place and a sense of purpose is far greater than can be quantified by an early retirement benefit. This sense of self-worth, accomplishment and connection with the land competes with the desire for a lasting legacy. No retirement benefit or government policy can compete with the sense of knowing and working a piece of land, seeing it shaped by your labour and decisions and being satisfied by a life well done. For many farmers the farm staying in the family is the desired legacy, whether the family farms it is insignificant. This type of legacy can be accomplished through inheritance, not subject to a business succession, making many farm businesses "last generation" farms.

References

Baker, J.R. and Epley, E. 2009. *Iowa Farmers Business and Transfer Plans* [online]. Available at: http://www.extension.iastate.edu/bfc/pubs/IA%20 Farm%20Business%20survey%20results.pdf

Barclay, E.M. 2006. *Farm Succession and Inheritance: Comparing Australian and International Research.* Report to the Rural Industries Research and Development Corporation: The Institute for Rural Futures, University of New England [online]. Available at: http://www.rirdc.gov.au/fullreports/hcc.html

Bika, Z. 2007. The territorial impact of the farmers' early retirement scheme. *Sociologa Ruralis*, 47(3), 246–272.

Chevan, A. and Korson, J.H. 1972. The widowed who live alone: an examination of social and demographic factors. *Social Forces*, 51(1), 45–53.

Fairweather, J. and Mulet-Marquis, S. 2009. Changes in the age of New Zealand farmers: problems for the future? *New Zealand Geographer*, 65(2), 118–125.

Foskey, R. 2002. *Older Farmers and Retirement.* Unpublished report to the Rural Industries Research and Development Corporation, Canberra, ACT.

Hoppe, R.A. 1996. Retired farm operators: who are they? *Rural Development Perspectives*, 11(2), 28–35.

Janke, J., Trechter, D. and Hadley, S. 2009. *Farm Succession Survey Report 2009.* Unpublished Survey Research Center Report 2009/10.

Kirkpatrick, J. 2006. *Unpublished Research of Farm Retirement and Succession Planning in Southwest Wisconsin.* Madison: University of Wisconsin Center for Dairy Profitability.

Kirkpatrick, J. 2007. Executive summary of farm retirement and succession planning in Southwest Wisconsin. *University of Wisconsin Extension FARM Team Focus newsletter.*

Kirkpatrick, J. 2009. *Farm Succession Research, Trends, and Programs in Wisconsin.* Paper presented to the Farm and Risk Management Education Conference, Reno, AZ, 31 March 2009.

Lobley, M., Baker, J. and Whitehead, I. 2010. Farm succession and retirement: some international comparisons. *Journal of Agriculture, Food Systems, and Community Development*, 1(1), 49–64.

Mishra, A.K., Durst, R.L. and El-Ostra, H.S. 2005. How do US farmers plan for retirement? *Amber Waves*, 3(2), 12–18.

Mishra, A.K., El-Osta, H.S. and Shaik, S. 2010. Succession decisions in US family farm businesses. *Western Agricultural Economics Association*, 35(1), 133–152.

Mishra, A.K., Morehart, M.J., El-Osta, H.S., Johnson, J.D. and Hopkins, J.W. 2002. *Income, Wealth, and Well-Being of Farm Operator Households*. Agriculture Economic Report No. 812. Washington, D.C.: USDA, Economic Research Service.

Pietoa, K., Väre, M. and Lansink, A.O. 2003. Timing and type of exit from farming: farmers' early retirement programs in Finland. *European Review of Agricultural Economics*, 30(1), 99–116.

Potter, C. and Lobley, M. 1992. Aging and succession on family farms: the impact of decision making and land use. *Sociologia Ruralis*, 32(2/3), 317–334.

Ryan, M. 1995. Early retirement for farmers. *The OECD Observer*, 194, 24.

Strauss, W. and Howe, N. 1992. *Generations: The History of America's Future, 1584 to 2069*. New York: Morrow.

Thenlin, A. and Holmberg, S. 2010. Farmers and retirement: a longitudinal cohort study. *Journal of Agromedicine*, 15(1), 38–46.

Troll, L.E. 1971. The family of later life: a decade review. *Journal of Marriage and the Family*, 33(2), 263–290.

Uchiyama, T., Lobley, M., Errington, A. and Yanagimura, S. 2008. Dimensions of intergenerational farm business transfers in Canada, England, the USA and Japan. *Japanese Journal of Rural Economics*, 10, 33–48.

Chapter 11
Business Continuance and Succession Planning: A New Zealand Perspective

Mandi McLeod

Family Farming in New Zealand

New Zealand is sometimes called 'the world's biggest farm', with an economy strongly based on agriculture and horticulture. In contrast to some other developed countries, the farming industry in New Zealand has been and remains of importance in the economy of the country. In the second half of the nineteenth century, Europeans settled in numbers, apparently to good effect, '[w]hen Europeans got land it was immediately turned to good account. The Europeans cultivated it, improved it and endeavoured to make something of it to keep themselves and their families' (McKenzie 1896: 302). Early settlers from Britain to New Zealand were encouraged to travel because 'the old country' had little or nothing for them. The opportunity to own land was a major motivation, away from the landlord-dominated, deferential farming frame of the homeland. Macaloon (1999: 211) discovers that the prime motivation for wealth accumulation from these businesses, 'was not for the individual, but for the family. Wills reinforce other sources in showing that the family was both the motivation and the instrument in creating a fortune'.

Since this time, agriculture has been of huge importance to a nation with considerable comparative advantage for pastoral agricultural production. The geographically isolated location of New Zealand may serve it well, in this regard, but distance from the market presents a number of challenges, requiring a certain collective discipline in terms of treatment of countries interested in importing its products. A number of specific factors, human and physical therefore combine to make the study of farm succession in New Zealand, an interesting and enlightening exercise. This chapter considers, first, the impact on farming in New Zealand of a major shift in policy in the mid-1980s, before discussing the key features of the mainstay of farming and the rural community, the family farm. Further consideration then focuses on the most popular options available to facilitate succession, and new entrants more broadly, before arriving at a number of key conclusions. At this point it is worth noting that the chapter draws on a number of sources including work undertaken by the author during a study, as Nuffield Scholar, entitled 'Family business continuance: a global perspective' (McLeod

2009), and ongoing consultancy experience as family business strategy consultant on issues of farm succession in New Zealand.

At the turn of the twentieth century, most people in New Zealand had a direct association with a primary industry. By 1997, this had shrunk to only 15 per cent, with less than 5 per cent working directly on the land (Rowarth and Caradus 1998).

Over the last decade more than one million hectares of New Zealand's countryside has moved from pastoral farming to alternatives such as grapes, olives, avocados, forestry and lifestyle blocks. However, meat, dairy products and wool remain among New Zealand's major overseas export earners, with dairying contributing around 22 per cent of the country's annual export income, where downstream processing is included (Federated Farmers of New Zealand 2011). According to Professor Rowarth in a recent article for Idealog, New Zealand's milking herd extends to '4.4 million dairy cows (approximately one for each Kiwi) and earns $10.4 billion in exports. As prices increase, the benefit is clear, each $1 increase in dairy payout is another $270 for every person in New Zealand' (Idealog website 2011).

Farming in New Zealand is characterised by the prevalence of family owned and operated farms, at the mercy of the international market and therefore the exchange rate, which dictate to a large extent the level of farm incomes in the country. In common with other developed countries, polarisation of size continues, with larger dairy herds of 500–1000 cows and sheep flocks of 3000 stock units and above, all chasing economies of scale in a competitive market (Parker 1998). In addition, land use changes from pastoral land to forestry or viticulture has an impact on the labour resource required, with consequences for family farm members. At the other end of the scale, smaller units have become a base for multifunctional activity – a lifestyle unit of the type that have proved to be attractive amongst city folk, as a means of escape (Paterson 2005).

Post 1984 Challenges to the Family Farm

New Zealand's experience of economic restructuring since the mid 1980's is unique. 'While by no means the only factor affecting New Zealand's agricultural sector, few would deny that the magnitude, pace, and impact of policy changes implemented by the fourth Labour Government from 1984 onwards represented the pivotal "radical break" from existing processes' (Johnsen 2004:419). Such policy changes included the abolition of farm subsidies which, in addition to concerns over fiscal and exchange problems, farmers had become increasingly critical of (Plank and Johnstone 1994). Literature abounds on the responses of the businesses to such challenges.

A review of the impact on the nation's farmers on a national scale conducted by Fairweather (1992) concluded that, 'there has not been a demise of family farming and a movement toward corporate farming. If anything, exposure to financial pressures has intensified the family character of farming'. Moreover, this study identified that the family farm had adapted with metamorphosis into a newer form of family farm, well positioned to deal with future challenges. Such adaptations were further investigated in a district to the north of Dunedin, South Island. This study identified, what others have referred to as 'dis-embedding' (Giddens 1991), or the weakening of 'the ideological linkage between hard work, land development and success' (Johnsen 2004: 427). Economic challenges to the business had invoked a number of responses including the reorganisation of labour and other resources, reducing reliance on farm income, broadening social networks encouraging the 'renorming' of self-perceptions of what it is to be a farmer. A 'new version' of family farming now prevails.

Farm Succession in New Zealand

In New Zealand, succession is a word loaded with different meanings to different people, with perhaps the most popular of these interpretations, the transfer of the family assets to one or more heirs. Most strategies designed to achieve this are based on a combination of tax minimisation and the division of the assets, in either equal portions or on the basis of gender. The question of fairness is seldom raised, nor is any attention paid to the dreams, desires and expectations of the people involved, until it is too late. More emphasis is paid to the value of the asset(s) than the overall viability of the business. Yet it is the family members that will determine the success or otherwise family business in the future. It can be argued, therefore, that the emphasis should be placed on the individual family members and how they function as 'a family in business' as this is more important to success or failure than the farm assets.

Keating and Little (1997) investigated this in a grounded-method study of family farms in NZ. Conducting interviews with first to fourth generation farmers (the latter a familial lineage on the farm back to 1885), the process of succession was identified as a sequence of *watching* (for signs of interest and ability), *reducing* (the number of potential successors, in terms of gender (male dominates), health, skills and availability), *assessing* (the skills and aptitude of the chosen successor, sometimes with a trial period) *compensating* (the non-successors, by payment for career training or purchase of a house, for example) and *placing* (handing over the business to the chosen successor).

An additional consideration in New Zealand, is the concept of 'OE' or overseas experience – an everyday expression in the country. Many see this as a 'rite of passage' for the young Pakeha (white) adult, with the opportunity to travel to distant places, away from the familiarity of home. This feature has been common in many farming families, where the son or daughter is encouraged to travel and work in countries overseas. The main benefits from this include a broadening of the mind, development of knowledge and skills relating to production, approaches to business and marketing, and a greater appreciation of the cultures and demands of other potential future markets. In addition, the OE participant develops a greater degree of independence and is, perhaps for the first time, able to understand the position of NZ in the world perhaps along with a greater appreciation of their own nationhood (Bell 2002). Tangible evidence of the effect of this 'pilgrimage' or 'rite of passage' is difficult to discern, although, anecdotally the benefits are substantial to the potential successor returning to take over the family farm.

Geographical isolation, physical distance to markets and the recent history of high land values relative to their productive capacity has, to some extent, driven the dairy industry in particular to consider alternative management systems and ownership structures to maintain viability and grow businesses. Traditionally, New Zealand has had a relatively self-replacing dairy industry through share-milking which allowed new entrants the opportunity to build up a herd of cows and develop crucial business skills before purchasing their first farm. Whilst this has allowed a ready-market of first farm buyers, most dairy farmers still aim to pass the family farm on to family members. The sheep and beef industry has a more traditional approach to succession, with family members taking over the family farm or the property being sold and proceeds divided equally amongst the family.

Succession Options – Share Milking, Equity Partnerships and Leasing

New Zealand has a number of different entry options for both new entrants to farming, with or without succession opportunities. Whilst many of these options have been developed by and for the dairy industry, many are now being adopted in the sheep and cattle sector, as part of their business continuance and succession planning.

The main options most relevant to existing farmers generally involve a business partnership of some variety, either as an income share or lease arrangement. Three approaches will be discussed, share milking, equity partnership and leasing.

Share-milking

Traditionally, share-farming, or share-milking as it is referred to in New Zealand, was developed as an opportunity for those wishing to purchase their first farm, to build up capital via livestock ownership and a strong cash flow. Today, approximately 40 per cent of New Zealand dairy farms operate a share-milking arrangement (Thompson 2011). Two main variants of share-milking operate in New Zealand – variable order (sometimes referred to as lower-order share-milking) and herd-owning share-milking. Variable order share-milking is where the landowner also owns the herd and the share-milker provides the management expertise and labour. Parties in this instance are covered by the Share-milking Agreements Order 2001, which specifies the terms and conditions negotiated by the parties. In herd owning share-milking, the share-milker provides the herd, management expertise, labour and chattels required to operate a dairy business. The parties are free to negotiate any terms they wish. In general terms, 'the contractual price' is agreed upon by both parties based upon the degree of management autonomy, relative investment in the business, expenses involved and the perceived risk involved in each situation. The price is expressed as a percentage of the total milk-solids produced. Agreements vary around the norm of 50/50 sharing, dependent on the relative input from the parties. The agreements are generally for a period of three years or more, to provide a reasonable period of security of tenure.

There is a high degree of interdependence between a farm-owner and share-farmer which encourages both parties to have regard for each other. As such arrangements invariably involve a shift in management control from the farm owner to the share-farmer both parties need to identify and agree on the level of control that will be assigned to each party which will in turn, dictate the profit share. A share-farmer with limited or no management control may be considered an employee irrespective of the agreement. Whilst the share-farmer is often responsible for employing and managing any labour required to successfully run the farms, it is the farm-owners responsibility to provide adequate housing for the share-farmer and their employees.

When both the farm owner and share-farmer work with integrity and respect, mutual benefit occurs, as both parties bring differing assets and skills to the joint venture and a mutual sharing of risks and benefits. Both parties become part of a larger system, contributing to progression in the industry. The arrangement assures a good level of rejuvenation in the dairy farming industry, by individuals who have, with determination and skill, proved themselves in both financial *and* operational management. Share-farmers are often, out of necessity, the innovators in on-farm improvements – in both information and technology, and the quickest to adopt new ideas. Their innovative approach and the preparedness to take risk and try new ideas, has meant that share-milkers often take on leadership roles in the industry.

Share-farming is seen as a good way to attract and retain good people in the dairy industry by offering the potential for personal and professional growth, asset creation and or realisation. In addition, it can provide a succession plan for many dairy farmers allowing them to manipulate the extent to which they desire to be involved in their business and how much capital they wish to release. In addition, it is an effective way of providing an incentive for share-farmers to stay for the whole season and for both parties to see the operation run at its full potential. The risks are shared and in time, 50/50 share-farmers often become the purchasers of farms, thus creating a market for non-family and family farm succession.

In summary, the strengths of the share-farming system are relevant to both parties. For the landowner, it offers the opportunity for the successors to take over the management of the business in a staged process, starting as a variable share-milker on a low share, building to high shares as time progresses. This releases the farm owner from the physical labour of the enterprise, whilst retaining a degree of input in the business, in terms of its management and future direction. In this way the arrangement provides an ideal training ground for successors or non-successors from outside the family.

For the successor or new entrant, the benefits are clear. The arrangement provides a step towards farm/business succession and an incentive for share-farmers to be rewarded for their skill and effort. There is also provision for share-farmers to be connected to the impact that their management decisions have on the business and therefore provides a skills development pathway. Perhaps most importantly, this provides the opportunity for the share-farmers to gather equity toward their ultimate goal, the take-over of the family farm or the purchase of an alternative holding.

In common with many other parts of the world, the value of farmland in New Zealand increased dramatically since the 1990s. Eves and Painter (2008) compared farmland returns from Australia, Canada, United States and New Zealand. They noted that, since 1990, the price of New Zealand farmland has averaged 40 times earnings. As a consequence, the share milking arrangement has come under pressure. In this context, Stevenson and O' Harrow, in an article for the Center for Integrated Agricultural Systems in 1999 (CIAS website), note, 'Share-milkers we interviewed told us that they need to own 350 to 600 cows debt-free to buy a first farm of 150 to 200 acres, but 15 years ago they would have needed one-third to one-half fewer cows to do so.' With rising land prices, the pressure has worsened. As if this were not enough, dairy cow prices have increased by an average of $500/cow in recent months, as a result of a developing market for breeding stock in China. There is talk of cow shortages in New Zealand in future seasons which will further increase, or at least hold, inflated cow prices. This compounds the difficulties for new share-milkers attempting to put herds of good quality cows together. The trend, identified by Stevenson and O' Harrow in 1999, continues, with New Zealand farmers, 'spending more time as share-milkers with increased herd sizes, before they take over or buy a farm' (CIAS website 1999). As an alternative, some individuals may consider an equity partnership arrangement, whilst others have decided that farm ownership is no longer an attainable goal. This does, however, decrease the pool of available first-farm buyers as individuals and increases the number of farms sold as syndicates or equity partnerships. Additionally, as the elements of the share farming equation change in value, different share agreements are expected. Until recently, this has meant a move for some landowners to requesting 60/40 arrangements, instead of the norm of 50/50. It will be interesting to see how the inflation in cow values further affects this balance. Stevenson and O'Harrow's closing observation in 1999 stands true today, 'Reduction in the number of share-milking opportunities and the increased economic pressure on the share-milker appear to be pushing New Zealand towards a significant decline in the rate of entry of new farmers' (CIAS website 1999).

Equity Partnerships

Equity partnerships are an increasingly popular response to the trend towards larger farms and the rising cost of land. For many, equity partnerships are a viable ownership structure that provides an opportunity to achieve business and personal goals that could not be achieved alone (National Bank 2008).

Equity partnerships exist between interested parties who jointly provide the equity to, either purchase a share in the property or the property in its entirety. They may be comprised of groups of investors, whose only relationship between each other is their investment in the property, groups of friends or family, or between a farm owner (individual, partnership or company ownership structure) and their manager or share-milker (often known as an equity manager/share-milker). Generally, those involved in equity partnerships have knowledge of the investment and some experience in the rural sector (Rabobank 2008).

For investors, joint venture or equity farms may present an opportunity to grow their business and achieve efficiencies of scale with shared capital outlay. The managing equity partner can benefit from ownership, with less requirement for capital input. For managers and share-milkers, equity partnerships can help them progress to full farm ownership, while retiring farmers can remain involved in the industry through their investment.

Equity partnerships offer shared capital, in addition to shared risk, and the leverage of specialist skills, and or capital, often results in the achievement of high performance and/or efficiencies of scale. They can also be an effective method of dealing with the issues of management and/or ownership succession. To deal with the issues of continuity, transferable ownership and limited liability, the most preferred ownership structure for an equity partnership is as a private company with a number of shareholders who are each issued shares, according to the amount of money they have invested. A company structure, in most cases, offers the benefits of transparency and regulation. It is reasonably common for one of these shareholders to manage the property on a day to day basis, either as salaried employee or as a share-milker, but this is not an imperative, nor is it any guarantee of success!

Where there is no prior history or relationship between the proposed partners, it is sometimes recommended (by the banks, accountants and other professional advisers) that planning sessions are held to explore common values and goals. With polarisation of farm size trends continuing, Equity partnerships offer an alternative vehicle for bringing a successor(s) into, what often are, larger agricultural businesses. They have many of the benefits of the share milking arrangement for the principal wanting to step back from the physical aspects of the business, and a clear and supportive structure for the successor entering a much larger business.

Leasing all or Part of the Land

Leasing maybe an attractive option where there is reluctance to part with the family farm and/or where there has been no decision on future options (investment, lifestyle or business) that could result from the proceeds of the farm sale.

In addition, leasing may a viable option for those starting out who may not have enough equity to purchase the farm outright or as a progressive process towards purchasing the farm. Anecdotal evidence suggests that family farms are leased to the next generation as part of the succession process, allowing the tenant to generate more income and build equity faster than if they were employees or share-farmers.

In general terms, a lease is the contractual agreement between a land owner and a tenant to lease the land and its fixtures such as sheds, fences, irrigation, water system etc. The lease agreement should cover the maintenance and ownership of these and ensure that the pastures and fertility of the property are maintained at least. Policies covering land use, maximum stock numbers and feed on hand at the end of the agreement should also be addressed. These agreements are legally binding and exist to protect both parties and should be clearly understood and agreed to by both parties prior to the tenancy. Provisions for dispute resolution process and exit clauses should be included. Determination of the rent payable, in New Zealand, will most likely be based on a range of factors, including the quality of the land, the amount of water available, the nature and condition of infrastructure and improvements and the quality of the house.

For the successor, a lease may provide a good degree of autonomy to develop their own business/enterprise, as part of the succession process. For the principal in the farm business, leasing can provide a flexible arrangement in the succession process, in terms of land included and term (length of lease). The attraction here is as a means of bringing the new generation onto the farm, whilst retaining ownership and control over the main family asset.

In addition, a rental income will be forthcoming, along with reduction in the burden of the day to day management of part or all of the holding. It may also allow retirement or semi-retirement before the successor is available, should they be on OE, still in formal education or working out an employment contract elsewhere.

Business Continuance – a Forgotten Aim?

An alternative view to conventional succession to property and business, is the move to greater *focus on the continuance of the family farm business* – with or without family members at the helm. This interpretation assumes that the business is in a position to be continued beyond the current generation and that this is the deliberate intention of the current owners. This model works on the principle that the current and future needs and expectations of relevant family members are identified and a plan put in place to effect this. Communication and conflict resolution skills are essential in ensuring that all are able to fully participate, if necessary, as business partners, in addition to being members of the family, and that time is set aside for business meetings, to ensure family time is just that.

Inherent in the business continuance model is the commitment of founders to passing the farming business on to another generation, *whether in the family or not*. If the founders are unable to make this commitment, the farm could become a personal 'old folks home', devoid of staff – 'a place to die with your boots on!' (McLeod 2009). In cases where a suitable successor cannot be found within the family, the same process is applied to ensure that succession of the business is successful, whether it remains in possession of the family or not.

Family business continuance and succession requires an effective process for determining the in-family heir (as opposed to an equal share amongst siblings) or non-family successor, which includes an allowance for 'sweat for equity' and/or the time-value of money. An analysis of the skills and knowledge required to operate and grow the business must correlate with the relevant current skills and ability of potential successors. Where possible and appropriate, a training plan is developed and effected or alternative solutions considered, which might include a supporting team, an external successor or sale. No-one wants to set their family up for failure by supporting them into a venture that is not, cannot and will not be successful, either due to their management capabilities (including business and financial acumen) or the scale and scope of the business.

Pessimism in the agricultural industry appears to be confusing the issue of business continuance with succession planning, the latter perceived as concerning the transfer of assets, at no real cost, with no debt incurred, to the in-coming heir. This can cause a predicament where the business assets are the only assets held by the family. The result can include the reduction in size of a business, which in most cases would have benefited from growing in size, or the loss of complementary elements of the business. In an international study of family businesses, by Grant Thornton Pty (2009), less than 10 per cent of family business owners are financially independent of their businesses when they retire. Anecdotal experience in the farming industry in New Zealand reveals that this holds true for family farming businesses and is possibly one of the main reasons that owners fear the process so much. Other anxieties arise through perceptions of the likely loss of identity 'as a farmer', loss of control, and an unwillingness to upset anyone (McLeod 2009).

However, many farmers 'don't know what they don't know' and without the processes to take them through a sustainable and systematic approach to continuing their business, this culture of pessimism and apathy will continue. What is required is a fair process that focuses on the family, the management and ownership, as distinct but inter-related identities in the business and allows for any conflicts in, or between, these identities to be resolved *prior* to any transition. Family businesses must be or have the prospect of becoming viable, sustainable entities, if they are to be viewed as businesses to be continued, vs. assets to be sold (Baker 2009). It is often the lack of knowledge concerning the business's current or future viability, against the value of the land asset, that prevents many New Zealand family farm businesses from initiating the process of business continuation.

In the traditional definition of succession in a family business, the intention is to pass on the assets to the next generation, based on either following primogeniture or as equal shares amongst siblings. Perhaps less thought is given to the 'business' *per se*, and the emotional complexities, inherent in a family business. In truth, these aspects may be the most significant in determining success or failure of the transfer.

A lack of transparency and ineffective communication between individual family members are cited amongst the main causes of conflict and failed management transfers. A recent study by Klinefelter and Voeller (forthcoming) reports that 60 per cent of failure in recently transferred businesses is attributed to these issues, with only 25 per cent as a result of poorly trained successors.

Business continuance is, therefore, a concept requiring action to ensure that not only the family connection but also *the business* succeeds through acknowledgement and implementation of the following:

- Orderly transference of responsibility for business management from one group of people to the other (Napier 2010).
- Relationship management and planning – managing the conflicts between family, management and ownership and asking the critical questions.
- Ownership vs. Management – what do the owners of the assets want/need in relation to the needs and expectations of the business?
- Asset management and protection – ensuring that 'current' assets are managed in such a way that they continue to grow and are not at risk from ill-conceived plans.
- Key people – family vs. non-family; what skills does the business need to continue and who has these or who can develop these through shared experiences and wisdom?
- Identification, attraction and retention of key skills required to ensure that the business has sustainable growth into the future.
- Wealth creation and/or preservation – ensuring that assets outside the business are created to prevent reliance on one asset to fulfil all future financial requirements.
- Financial structures for financial freedom – having the right financial structures in place that best fit the strategic direction of the business and are fiscally sound and up to date from a financial efficiency perspective.
- Estate and retirement planning – plans in place to ensure that there is independent income and leisure activities to prevent post-farming depression and create feelings of self-worth independent to the farm.
- Strategic Planning – everyone should know, understand and take ownership of the implementation of the strategic direction the business must take to meet the financial, social and emotional goals of the owners and key stakeholders. This vision should be jointly held between the owners and key stake holders who have all participated in its creation.
- Growth or death – identification of where the business is on the business life cycle is vital in establishing the strategic direction and growth plan.

With any shift in control in a business, there is potential for conflict. In family farming the ones to watch are based around the business values and philosophy, in terms of how the farm business will operate and the strategy for decision making. Inevitably, conflict is caused when there are unmet expectations and a lack of clarity around the roles of family, business management and ownership (Grant Thornton Pty 2009).

Kiwi Ingenuity and Commitment in Family Farming – Future Perfect?

Families are complex, consisting of individuals linked through a common lineage, sometimes with little more in common than genes. It is the individual that forms the heart and soul of a family-farm, ensuring its ultimate success or failure, yet it is these same people that are forgotten about in a traditional succession planning model, which is developed around tax-effective mechanisms for reallocation of assets when the matriarch and or patriarch dies.

The average age of farmers in New Zealand is 58 years (Baker 2008), and, as a consequence, management and financial decisions are in the hands of a generation who may be more resistant to change and growth. This, coupled with a shortage of professional family farm business specialists, places New Zealand family farms in a difficult position, where technological development and information transfer will undoubtedly determine future on-farm profitability and infrastructure development.

Changes to the entry cost of farming, relative to income, is altering the perception of farming (increase in land and cow value; introduction of Fair Value Shares – each shareholder supplier must hold shares in Fonterra relating to their productive capacity). The challenge for the next generation is great, as they struggle to prove viability to meet increasingly tight financial lending requirements. Unlike their European counterparts, there are few non-agricultural opportunities for increasing incomes through farm shops, barn-conversions into office and retail space, or even through farming for environmental stewardship revenue. As more of the 'instant-generation' are put off by the long hours and low incomes, the questions of 'who' will feed the next generation is increasingly problematic.

How to attract new entrants to agriculture, especially at this time of public 'scrutiny and challenge' over the environment, animal welfare, food and energy security and supply issues, remains central to the issue of succession and farm business continuance. If there is no-one to take over the family farm, no entrants willing to purchase farms, the flow-on effects will be read in land prices and asset valuations. Rural lending institutes, accountancy firms, industry sector groups and Federated Farmers of New Zealand are presently focused on this complex issue.

Historically, New Zealand farmers have been able to treat their business as an extension of their lifestyles, sometimes fostering less developed business management skills. However, today's farmers must embrace their operation as a profitable and sustainable business, adopt practices long since adopted by their urban-peers; that is accept full responsibility for financial management and control and a desire to see their business independently from their own psyches and not as a job. Therefore, the business should be capable of existing beyond the founder's passion and 'traded' between generations of successors, whether family or not.

On its own, and with the previously identified connotations attached, the word 'succession' is no longer appropriate for the processes that family farm businesses need to go through, if longevity of their business is the key outcome desired. Possibly the term 'business continuance and succession planning' emphasizes that the focus must be on people and business management, not property and asset transfer. In New Zealand, family farms need to grow or they will die as a result of a lack of investment.

The future of farming must be seen as important and valued with a clear vision. People will not do what they don't want to do – what they don't value or deem to be important. It is therefore imperative that the farming community sell themselves as worthy business managersm with a business that has an intrinsic value beyond that of the assets involved.

A program of business continuation for the farming sector should be embraced by farmers, industry and governments, who must surely recognise the importance and contribution of agriculture to their own economy in addition to global food supplies. Such a program is required to take into consideration the generational differences between the current farm owners and the new generation – the 'instant gratification' generation; the fear vs. opportunity that can arise out of a well-managed transition; time management and not time drift; reinvention and not retirement.

Whilst the New Zealand dairy industry has developed programs such as share-milking and contract milking, which have historically allowed an ease of access for young farmers to enter the industry and progress through to farm ownership, changes to regulations around the cost of supply to Fonterra[1], for example, and the dramatic increase in land prices have made this progression more difficult. The increased capital requirements have largely predicated alternative models for entry, such as the implementation of Equity Partnerships. In turn, this has allowed farmers to increase their holdings through strategic partnerships with investors, making it more challenging for some who wish to get into farm ownership in their own right. For others, Equity Partnerships may offer new opportunities to obtain a stake in ownership and gain experience, advice and mentorship from the other equity partners.

There is no doubt that farming in New Zealand has advantages over many other parts of the, although it does perhaps have a narrower range of opportunity in terms of multifunctionality. New Zealand farmers are renowned for their ingenuity, and rewards will go to those who see the possibilities before they arrive. New forms

1 Fonterra is New Zealand's largest farmer-owned dairy co-operative, which handles 90 per cent of all milk produced in New Zealand.

of farm succession will continue, with the focus very definitely on scale and the business of agriculture.

References

ABARE 2009. *Outlook 2009*. Available: www.abare.gov.au/publications_html [accessed 26th June 2011].

Baker, J. 2008. *Your Last Fencepost: Succession and Retirement Planning for New Zealand Farmers*. Longacre Press, Dunedin.

Baker, J.R. 2009. Personal communication.

Bell, C. 2002. The big OE: New Zealand travelers as secular pilgrims. *Tourism Studies*, 2, 143.

CIAS website (Center for Integrated Agricultural Systems) 1999. *Sharemilking in Wisconsin – Evaluating a Farm Entry/Exit Strategy*. Available at: http://www.cias.wisc.edu/future-of-farming/sharemilking-in-wisconsin-evaluating-a-farm-entryexit-strategy [accessed 28th July 2011].

Eves, C. and Painter, M. 2008. A comparison of farmland returns in Australia, Canada, New Zealand and United States. *Australian and New Zealand Property Journal*, 1(7).

Federated Farmers of New Zealand 2011. *Farming Facts*. Available at: http://www.fedfarm.org.nz/about_us/farmingfacts [accessed 27th July 2011].

Giddens, A. 1991. *Modernity and Self-identity: Self and Society in the Modern Age*. Polity Press, Cambridge.

Grant Thornton Pty 2009. *Succeeding at Succession*. Available at: www.gti.org/files/phb_succeeding_at_succession_overview.pdf [accessed 1st July 2011].

Idealog 2011. *Why our Future Lies in NZ Food Inc.* Available at: http://idealog.co.nz/blog/2011/07/why-our-future-lies-nz-food-inc?utm_source [accessed 28th July 2011].

Investment NZ *New Zealand Facts and Figures*. Available at: http://www.investmentnz.govt.nz/section/15291.aspx [accessed 27th July 2011].

Johnsen, S. 2004. The redefinition of family farming: agricultural restructuring and farm adjustment in Waihemo. *New Zealand Journal of Rural Studies*. 20.

Klinefelter, A. and Voeller, M. (Forthcoming publication). Successor Development and Management Transition on Family Farms and Ranches. To be published in *Top Producer*.

Lindsay, R. and Gleeson, T. 1997. *Changing Structure of Farming*. Available at: http://adl.brs.gov.au/data/warehouse/pe_abarebrs99000533/PR11684.pdf [accessed 25th June 2011].

McKenzie, J. 1896. Minister of Lands, *New Zealand Parliamentary Debate*. 92: 302.

McLeod, M. 2009. *Family Business Continuance: A Global Perspective*. Report to the Nuffield Foundation.

Macaloon, J. 1999. Family, wealth and inheritance in a settler society: the South Island of New Zealand 1865–*c.* 1930. *Journal of Historical Geography*, 25(2), 201–215.

Napier, R. 2010. Personal communication, Napier Agrifutures.

National Bank 2008. *Equity Partnerships*. Available at: http://www.nationalbank.co.nz/rural/specialistservices/equity.aspx [accessed 28th July 2008].

Parker, W. 1998. Preparing farm management for the new millenium. *Primary Industry Management*, 1(2).

Paterson, J. 2005. *What is a 'Lifestyle Block' and is it a Form of 'Rural Gentrification'?* Paper presented to 'Focus on Rural Research', Waikato Branch of the NZ Geographical Society, Hamilton, 17 November 2005.

Plank, R.D. 1994. *Farming in a Subsidy-free Zone*. Paper to the American Agricultural Economic Association Conference, San Diego, California.

Rabobank, 2008. *Equity Partnerships*. Available at: www.rabobank.co.nz/Rural/../Equity-Partnership-Facilitation.aspx [accessed 26th July 2011].

Rowarth, J.R. and Caradus, J. 1998. Primary production professional beyond 2000. *Primary Industry Management*. 1(2).

Thompson, A. 2011. *Share Milking*. Available at: http://www.fedfarm.org.nz/n274.html [accessed 28 July 2011].

Chapter 12
Succession Planning in Family Businesses: Consulting and Academic Perspectives

Peter C. Leach

Introduction

Sir John Betjeman, England's Poet Laureate from 1972 to 1984, was the unlikely victim of the nasty things that can happen when family business succession goes wrong. The Betjeman's family business was cabinet-making, not farming, but the particular commercial activity is not crucial to the discussion in this chapter. The family's experiences highlight a number of the challenges that disrupt and sometimes destroy the successful transition from one generation to the next of otherwise sound and prosperous family businesses.

In his autobiographical poem, *Summoned by Bells,* Sir John records how he turned down his father's calls for him to take over control of the family's third generation business. Despite John's decision early in childhood that he would become a poet, his father Ernest never gave up pressuring his only child to join the company, and each time John, with growing feelings of guilt, exasperation and self-reproach, had to refuse his father's pleading.

Eventually the campaign of emotional blackmail saw father and son divided and estranged and, some ten years after Ernest's death in 1934, the business he and his forebears had built up was liquidated. The clear irony is that it's hard to imagine anyone less suited to a business career, let alone family business leadership, than John Betjeman – the firm would likely have found itself insolvent much sooner with him at the helm. But, as we will discover, realities like this have no bearing on the guilt endured by a next generation family member, encouraged to feel they are to blame for letting down parents, family tradition, the family business and the firm's loyal employees.

In this chapter I'll focus on strategies to help avoid this sort of disaster, based on putting in place systematic and well-structured plans for succession – the most serious long-term challenge that family businesses face. This book so far has provided many excellent perspectives on family farm succession and retirement (a word I worry about because of the negative connotations it still carries!),[1] but

1 The words 'retire' and 'retirement' inevitably crop up throughout this chapter and, although the dictionary definitions of these words are in transition, they still tend to be associated with unhelpful concepts involving an end to the productive period of someone's

here there is a chance to take a step back and to sharpen the focus on the nature of 'family business' and, as a starting point, to review the key lessons learned from conceptual thinking, research and analysis.

Family Businesses – Vital, Special and Complex

Family businesses power innovation and enterprise around the world. In the United States 90 per cent of businesses are family businesses, accounting for 60 per cent of employment (Okoroafo and Koh 2009). In the UK they account for 65 per cent of the economy's 4.5 million private sector enterprises and provide jobs for 9.5 million people – one job in three (Institute for Family Business 2008). Yet too often, family businesses are grouped together with the small and medium-sized enterprise sector, unhelpfully disguising their exceptional status and significance.

Unique Dynamics

It has long been acknowledged that family businesses are special, benefiting from a range of advantages not found in other enterprises. Family involvement promotes long-term orientation, patience, common values, strong commitment and the idea of 'stewardship' – that is, responsibly managing entrusted resources so as to hand them on in better condition. (The latter idea is encapsulated in the philosophy of France's Hermès family, still in control of the fashion house established by Thierry Hermès in 1837: 'We do not inherit the business from our parents, we borrow it from our children.')

Despite this special status, however, it is only in the past 30 years that academics and commentators have begun to treat family businesses as an independent study discipline. Slowly but surely across this period the sector has become a vibrant area of interest for researchers, theorists and commentators. A key conclusion of the research effort – that family life and business life do not rest easily together, and that there is a culture clash – may sound obvious, but it was not obvious at all before the thought was formulated! Today, this powerful insight about complexity has become the foundation of a logical framework helping us to visualise what is going on in family businesses, and to understand why running them, and successfully passing them on to the next generation, can be such hard work.

Figure 12.1 shows that family life is emotion-based with its members bound together by emotional ties; it tends to be inward-looking, placing high values on long-term loyalty, care and the nurturing of family members; and lots of family behaviour is influenced by the subconscious – for example, the need for parents to treat children equally, and the need for fathers to be stronger than their

---

life. The main point as regards this chapter is that the challenge for family business leaders has less to do with deciding how to leave their business than working out how to reshape their connection and attachment to it.

sons. Business life, in contrast, is based on the accomplishment of tasks, with the emphasis on performance and results; it is outward-looking, built around contractual relationships in which people do agreed jobs in return for agreed remuneration; and, for the most part, behaviour is consciously determined.

In family businesses these radically different, essentially incompatible, cultural domains not only overlap, they are actually interdependent. Their differing purposes and priorities produce the special tensions that exist in family firms, creating at the overlap points operational friction and value conflicts for owners and other family members.

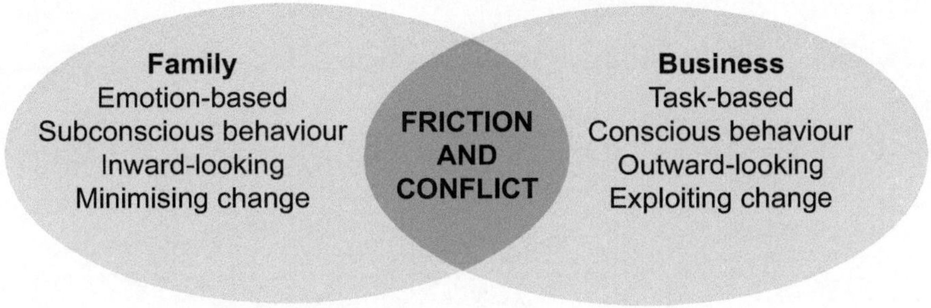

Figure 12.1 Overlapping Systems
Source: Leach (2011)

Controlling the Culture Clash

Difficulties arising from the overlap of family and business domains cannot be avoided entirely, but successful families devise strategies that help them to keep the overlap under control. A key component of these strategies is developing the ability to become an observer of one's family business as a 'system'. Under a systems perspective the family firm can be modelled as comprising three overlapping, interdependent sub-systems (see Figure 12.2), and this mental model helps us cope with the idea of three lots of things going on at the same time – change in the family, change in the business and change in the ownership profile. It is the latter that is the most critical variable in the present discussion, because understanding the ownership structure in a family business is fundamental to understanding the real forces that are at work during succession transitions.

This three-circle model offers the chance to focus on the complexities of family companies by identifying the motivation, expectations and fears of individuals within the three groups. (Everyone involved in a family business falls within one – and only one – of the seven sectors created by the three circles.) The model also highlights potential sources for interpersonal conflict and role confusion, emphasising how these issues are built into the very fabric of family businesses because the different stakeholders see the world differently and want different things from the business. For instance, when forming a view on the appropriate level of dividend payments, non-owning family members who work in the business (sector 6 in the diagram) often take a very different view from their relatives who own shares in the firm but are not employed by it (sector 4).

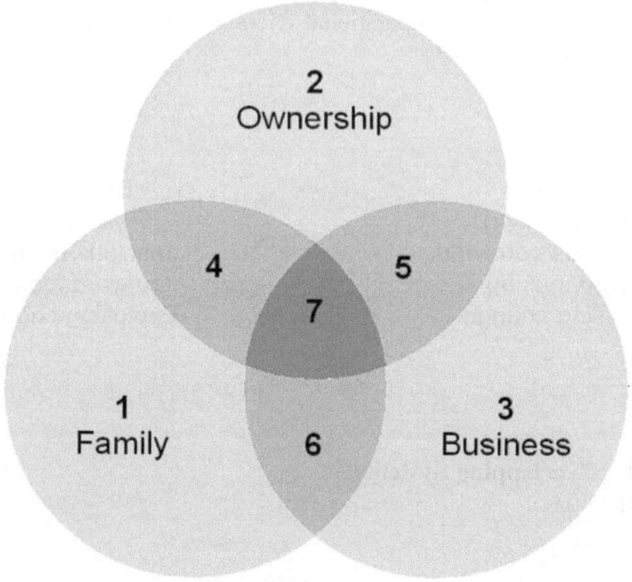

Figure 12.2 The Three-circle Model

Note: Key: 1. Family members; 2. Employees; 3. Shareholders; 4. Family shareholders; 5. Employee shareholders; 6. Family working in the business; 7. Family shareholders working in the business.

Source: Adapted from Tagiuri and Davis (1982)

Family business people need to develop special skills that help them to identify and manage the complications that these dynamics introduce, and to adopt methodical strategies to foster growth of the business and the transfer of power and control within it. This is important not least because the mortality statistics make such bleak reading. Worldwide, only about one-third of businesses successfully make the transition from each generation to the next, and some studies suggest that only around 5 per cent of family firms are still creating shareholder value beyond the third generation (*The Economist* 2004).

Resistance to Succession Planning

Succession confronts family business owners with a straightforward set of strategic options. They might plan for a family member to succeed them, or a caretaker manager, or a professional manager, or they might consider exiting via a sale of the business. A standard SWOT analysis can be applied to assess the strengths, weaknesses, opportunities and threats of the options for a particular family business.

Where the aim is to retain direct control over the business (and it's these cases that are our main concern here), seeking a 'family solution' – that is appointing a family member to succeed – is seen as particularly attractive by many owners. They feel their personal ideas and values will have a greater chance of survival, and that that their sacrifices building up the business will have been worthwhile, passing on to the new generation the special opportunities of family business ownership and leadership.

But there's another option that is missing from the SWOT analysis list above – adopting a 'do nothing' strategy – and it is here that we encounter the central paradox of succession. Doing nothing is the least logical, the most costly, the most destructive of all the options – yet it is the most popular. Failure to address succession is often put down to a combination of the entrepreneur's instinctive desire to keep control of his creation, as well as a natural aversion to planning. But the reasons are normally more subtle, and in many cases are rationalisations designed to avoid deep-rooted anxieties and fears. Understanding these forces is an important first step in successfully managing the transition process. Here are just a few examples:[2]

- *Fear of mortality*: often a particular problem for entrepreneurs, whose success is usually driven by a powerful ego and the conviction that they control their own destinies.
- *Loss of identity*: owners often identify strongly with the business, seeing it as a personal achievement that defines their place in the world.
- *Inability to choose among children*: under business principles, the choice of a successor should be based on competence, but family values, which dictate that children should be loved and treated equally, can prevail.
- *Spouses' resistance to change*: the owner's spouse is frequently reluctant to welcome and encourage a partner's move into retirement.

2 The US organisational behaviour expert, Professor Ivan Lansberg, was the first researcher to study this area, and his early work remains the starting point for further reading. He categorised the factors conspiring against succession planning into those connected with the owner, the family, the employees and the general environment in which the business operates: see Lansberg (1988).

- *Fear of parental death*: the next generation typically has deep-rooted psychological worries about abandonment and separation, and such feelings can be too painful to permit participation in discussions about succession.
- *External worries about change*: outside the company, important customers are also likely to prove resistant to change, reluctant to trust a new face.

So owners have to face up to a range of complex and interrelated processes – psychological, emotional, individual, organisational and external – that are all operating against any kind of planned effort to manage succession.

Grasping the Nettle

A key strategy to help overcome the forces favouring doing nothing about succession is for owners to prepare themselves financially and emotionally for a new phase of their lives that does not revolve around the family business. In summary, they need to seek security in four areas (Aronoff 2003):

- *Personal financial security* ('Do my spouse and I have the resources to live comfortably for the rest of our lives?').
- *Family security* ('I'm afraid the family will fall apart if I relinquish my leadership position.').
- *Organisational security* ('Can the business function without me?').
- *Psychological security* ('If I'm not the leader of this company, who am I? And what am I going to do?').

Financial Peace of Mind

Financial security is important for the family business owners' retirement, yet many owners neglect their personal finances. They often believe that the business (usually in ways unspecified) itself represents their personal nest-egg. This may well turn out to be true, but it involves a number of assumptions that cannot always be taken for granted.

Seniors therefore need a secure source of income that is preferably separate from the family business, to provide post-retirement independence and confidence. Broadly, financial peace of mind in retirement can either be achieved by business owners continuously taking money out of the business during their period of tenure, or leaving this money to build in the balance sheet so that, on retirement, a restructuring can be arranged to transfer personal wealth to the departing owner. Both strategies have pros and cons. It is possible to reduce or delay tax liabilities when the business is passed on, but this requires careful planning and professional advice.

Owners often do not have a large enough estate to enable them to leave the business to active, interested offspring, while leaving other assets to the remaining children. However, structures are available – including life insurance, hiving off property assets and isolating equity share voting control – that help achieve a compromise between short-term 'fairness' and solutions that are in the long-term interests of the business.

Psychological Ties – Preparing for a New Life

Emotional disengagement is most likely to be successfully negotiated if business owners are retiring *to* a new life of interesting activities, rather than from their old one, which implies that their useful and productive days are over. Owners severing their connection with the family business is neither desirable nor in fact possible – after all, the business is part of the fabric of the family. Owners must think, therefore, about how best to reshape their attachment to the business, and to plan their future work activities (Leach 2011, Davidow 2006). Bear in mind also that seniors can remain an important resource to the family business (as ambassadors and mentors, for example), even though they have passed on day-to-day operational responsibility to their successors.

Succession is much more likely to proceed smoothly if owners step down at a time when they are still in full command of their abilities and able to provide guidance to senior managers when they seek it. Life-cycle analysts have highlighted the difficulties that can occur when founders hold on too long and become 'out of synch' with their successors.[3] And as society ages, these pressures and imbalances look set to increase.[4]

Family business leaders who enjoy most success in passing on the business tend to start exploring disengagement very early on. A recent UK example was Christopher Oughtred, who began planning his exit as fifth generation chairman of William Jackson and Son bakers when he was in his mid-40s. He created a five-year plan during which he set milestones and targets for key aspects of his succession solution. The next generation were assessed and prepared, the business streamlined, and he developed a post-succession plan for himself and his wife, as well as his 'next career'.[5]

---

3 See, for example, (Hudson 1999) who defines the developmental challenges for people in their 50s and 60s as securing self-renewal and creating new beginnings, while, in contrast, developmental work for those in their 70s and beyond focuses on reflection, creating a legacy and mentoring. It is clear how family business seniors who are unable to start planning the renewal of their lives in their 50s and 60s can end up seriously 'out of phase' during their 70s and beyond, which in turn affects the lives of next generation successors, themselves in mid-life and probably impatient for independence, recognition and opportunities for leadership.

4 For a discussion of how demographic changes may impact on succession see (Murray 2005).

5 For the full story, see (Murray 2009).

Owners must decide when to step down and stick to it. Vagueness and imprecision about how long a family CEO or MD will run the company belongs to the past. There's a very strong argument for including written provisions in the family constitution and in the succession plan, discussed below, specifying how long individuals will be allowed to hold the top jobs.

Leading a Successful Transition

Once business owners have achieved a comfort zone with personal financial security and psychological security, they can move on to address the other two issues introduced in the previous section – family and organisational security – with the leader's goal to create a family business that can sustain itself when they have departed the scene.

Strong leadership is required to guide the business through the transition. Succession planning – done well – takes a long time, and the most successful transitions result from establishing a partnership process between the generations based on mutual responsibility, respect and commitment. They also result from a willingness to address some fundamental questions that otherwise risk staying firmly under the carpet, like 'Does the business need to remain family owned?', and 'What would happen, and who would take over if our family business leader dropped dead tomorrow?' An important aim of succession planning is to eradicate vagueness, assumptions or guesswork.

Once it is decided that the long-term objective is to keep the business in the family, here is my checklist of guidelines for managing the succession process:

- Start planning early
- Discover and manage expectations
- Encourage inter-generational teamwork
- Develop a *written* succession plan
- Involve family and colleagues, and seek outside help
- Establish an education and development plan
- Take the opportunity to renew the family's values and vision.

Start Planning Early

The problems family businesses face are mostly predictable, and it is important to take advantage of this helpful fact. Rather than waiting till the reading of the will to resolve questions like 'Who gets the shares?' or 'Who is best suited to take on managerial leadership?', in a family business you have the opportunity to address such issues ahead of time, in a calm atmosphere, under an agreed process, thus reducing the potentially damaging impact of unexpected yet predictable events.

Succession is a process, not an event, and it should be carefully planned and take place over time. Ideally, the senior generation's reduced involvement is so gradual as to be almost imperceptible. Current thinking (for example Aronoff, McClure and Ward 2003) recommends a period of between 5 and 15 years, with owners beginning to think seriously about succession at 45 to 50 years of age, and planning to retire at 60 to 65. Typically, when this phase is started, the next generation would be 25 or 30 years old, with their formal education and outside work experience behind them. Beginning the succession process at this stage allows a period of around a decade, which is ideal when choosing from among multiple candidates, and it also facilitates the development and grooming of potential successors, giving them ample opportunity to grow into their roles, earning the respect and confidence of the senior generation and other stakeholders.

Discover and Manage Expectations

A key conclusion from my consulting experience is that assumptions should never be made about what individual family members will want or expect in any given situation, and open communication and transparency must take the place of secrecy.

Setting up a family council in which family members can address how the business affects them and how they as a family can influence the business has been found to be one of the most powerful ingredients in successful generational transitions. Councils help nurture family cohesion as a business passes from one generation to the next: possible succession scenarios need to be evaluated and their viability measured against the dreams, talents and capabilities of the participants. It is a process of testing, learning and revision, and managing this exploration phase is probably the most important leadership challenge of transitions.

Encourage Inter-generational Teamwork

Establishing and fostering inter-generational teamwork is crucial if the build-up to the transition, the transition itself and what happens afterwards are to be as trouble-free as possible. There is no substitute here for leadership by the senior generation who, if possible, should become coach and mentor to the next, leading to a staged shift of power and control over time. Father–son and other family rivalries often inhibit the development of such a scenario, but such inter-generational partnerships, when they work, are powerful and effective.

With succeeding generations, the intensity of emotional factors that surround the family's involvement in the business can increase, and bits of 'emotional baggage' – unresolved issues left over from the previous generation – can develop a life of their own. Grudges and feuds may live on and fester in families for a very long time. Establishing inter-generational teamwork implies the generations recognising these sorts of problems, talking about them and trying to find solutions. Assuming such issues will just be forgotten about over time can be a fundamental mistake; the problems generally resurface later on, more complicated and more difficult to resolve.

Ensuring financial independence between generations is important and, if necessary, provision should be made for it in the succession plan.

Develop a Written Plan

Establishing formal mechanisms, rules and procedures help families to avoid (or at least manage) tensions and divisions, which, if left unchecked, interfere with the effective functioning of the business. Drawing up a written family constitution (recording the family's agreed policies on the business and other issues) provides a structural framework that helps family members focus on what's important, progress through problems and find ways of working with each other. And, for exactly the same reasons, developing a written succession plan that incorporates a step-by-step approach to the practical and psychological aspects of the transition process will prove invaluable.

The thought required to formulate and write down each stage of the succession process will be useful in itself, and the existence of a formal document that everybody is aware of, and has been consulted about, will significantly reduce the potential for doubts and misunderstandings. The timetabled plan should plot each phase of the senior generation's reduced participation in the business and make clear the process for choosing a successor. The plan should also lay down how its implementation will be confirmed and communicated to the family, the company and outside stakeholders (like creditors and key customers).

Involve Family and Colleagues, and Seek Outside Help

Ideally appoint a 'broad Church' succession working party, including seniors, selected family members, non-executive directors, key trusted employees and neutral outside experts. The group will be responsible for developing the succession plan and monitoring its implementation.

It's the business owner's responsibility to initiate and lead the succession planning process, but the working party gives everyone most directly concerned an outlet to discuss their thoughts and fears. By providing a forum for debate, it should help to reduce negative emotional reaction within the family.

Establish an Education and Development Plan

Many owners assume that their children will want to enter the family business, or (like Ernest Betjeman) they put pressure on them to do so. Inadequate preparation and training, or undue pressure, condemns many next generation members to unhappy careers that are neither satisfying for them nor productive for the business.

As discussed below, a programme to foster the personal development of potential successors should be initiated early and be made part of the succession plan, enabling evaluation of their abilities and – looking ahead – their suitability for leadership.

Renew the Family's Values and Vision

Last but certainly not least, succession should be seen as an opportunity to refresh the family's values (what a family and its business stand for) and vision (the shared sense of where each are headed). Values and vision provide a major source of strength and resilience for the family firm, and family businesses can achieve competitive advantage through a values-driven approach. But if values and attitudes remain static and entrenched in the past, the family risks creating a vacuum in which – with no relevant vision to unite them – disconnection, communication failures and conflict are more likely to flourish (Zalman 2005).

Renewal of a coherent ownership vision is required with each generational transition of the family business, and the flux of the transition process provides an opportune time for the family to collectively engage in this task before policies and positions take root.

Developing Effective Successors

Nurturing next generation leadership does not come easily to most entrepreneurs, but successor development is critical if the business is to be sustained and revitalised.

The key lessons I have learned from my family business consultancy career are that succession should involve a well-planned partnership with the next generation, and parents must take responsibility particularly for ensuring their children receive a sound, broadly based education; that they are well nurtured and develop self-esteem; that they learn about money, business and investment; that they have extended outside work experience before joining the family business; and that if they do join, there is an education and development programme for them that is both relevant and worthwhile.

To Join or not to Join?

While the children are growing up it is important to try and keep an open mind about whether they will want to enter the family business, and to remember that their perception of the business is being formed mainly on the basis of what the senior generation tells them about it. If they regularly hear complaints about the problems of running the firm, or on the other hand are conditioned from birth to believe that the business is a golden inheritance, this will colour their approach.

Owners should try to find a balance that enables their successors to share their dream, making sure excessive pressure is not put on them to feel that they have no choice but to be part of it. Let the next generation know they will be welcomed into the business if this is their choice and provided they have the necessary capabilities. But, equally, make clear they will be supported if they choose other careers.

Training and Development

A programme fostering the personal development of potential successors – acknowledging skills and attributes already possessed and listing competencies that need developing – should be part of the succession plan. Once formal education is complete, next generation family members should be encouraged to pursue other careers and work opportunities before joining the family company. This helps build self-esteem and confidence, widens business experience and will improve their credibility in the eyes of non-family employees. The most accomplished successor-generation managers of family firms have generally spent much of their early working lives outside the family business.

Once they have joined the business, make sure that all training is worthwhile and appropriate in relation to the career development strategy set out in the written succession plan. Also, be aware that because of the emotional involvement, parents can be poor teachers, so if possible help young family members joining the firm to establish a special relationship with a non-family mentor figure (often a senior manager) within the organisation. Defining next generation roles carefully, setting objectives and providing feedback are important. Conflict and uncertainty over their functions in the business can be a major source of tension. Best results are generally achieved by setting up tailor-made programmes for developing the next generation.[6]

Evaluating Candidates and Making the Choice

In evaluating candidates for succession, families should look with honesty at three key factors: (a) do they have the intellectual capacity and the organisational and people skills required; (b) do they have a real desire and commitment to take the reins and drive the business forward; and (c) while laying claim to leadership, do they have the support and respect of both the family and other stakeholders? Solid qualifications under all three of these headings are required.

Sometimes, the choice of successor seems clear cut. There can be a single successor – regarded by the family as the 'logical' choice – who is both capable and committed and who, during the succession planning process, grows naturally into the role. But some families define 'logical' to mean that the eldest son is automatically the first choice – a concept that is no longer sustainable. Although it eliminates uncertainty and reduces the likelihood of rivalry among the children, such a rule may result in the appointment of a leader who is less qualified and less

6 See, for example, (Webb 2005), and also the work of IMD family business professor, Joachim Schwass – in particular (Schwass 2005) – drawing together some key conclusions on effective successor development strategies.

competent than other candidates. Studies have concluded that family businesses run by eldest sons tend to be managed relatively poorly.[7]

Searching for a single successor in part reflects a bias in favour of the single-leadership model for family companies. There may be compelling reasons for family firms to combine two or more people in leadership, such as parental refusal to face up to the decision of choosing a successor from among their children, or it being too early to undertake a complete handover between generations (Nicholson and Björnberg 2005). Partnership models of leadership can work in a family business, but they depend on there being a degree of shared vision, clarity of roles and responsibilities, appreciation of diversity in personalities, and a strong bond that fosters collaboration and successful teamwork. It also helps if co-leaders have relatively equal abilities, plus a willingness to compromise and to accept consensus decision making. Even then, mechanisms should be in place for resolving deadlock.

Succession decisions based on gender issues are increasingly recognised as outmoded and absurd. Daughters often possess greater qualifications for the leader's job, demonstrated by strong early career achievement. And the absence of the potentially troublesome father–son relationship can boost the effectiveness of inter-generational teamwork and smooth a daughter's rise to power.

The involvement of committed sons-in-law or daughters-in-law can also provide a pool of potential next-generation leaders and bring new dimensions of strength to the family business, but the common occurrence of divorce can have a negative impact when it involves an in-law who is in a key management position. Families can anticipate such problems through pre- or post-nuptial agreements, but ultimately the risks of a potential marriage breakdown have to be weighed against the benefits an in-law can bring to the business.

Although family business leaders are responsible for initiating and overseeing the succession process, when it comes to making the choice some experts have concluded that, because of emotional and psychological factors, senior family members working in the business should not be responsible for selecting their own successors.[8] To avoid these dangers, the advice and assistance of a strong board of directors is invaluable, both in assessing the capabilities of family members in the business and any non-family candidates, and in making the final decision.

An interesting alternative approach to the 'Who should do the choosing?' question has been built into the family constitution of an Indian family business I have worked with. Five years before the owner's pre-set retirement date a process

7 For example, (Dorgan, Dowdy and Rippin 2006) concluded that the prevalence of such companies in France and the United Kingdom seems to account for much of the gap in effectiveness – and perhaps in performance – observed relative to family businesses in Germany and the United States.

8 See, for example, the early but still illuminating work of Harry Levinson (1974).

will begin to select and appoint a successor, and the constitution provides that it will be the next generation, forming a selection committee, who do the choosing.[9]

What if No One Fits the Bill?

Although owners may have a huge amount of emotional capital invested in the family business, and wish to see it continued by their children, it may be self-defeating to force a family management transition if the right circumstances do not exist. If, after an honest assessment, the conclusion is that there's little chance of a successful transition to the next generation, owners should begin to look for alternatives.

If the aim is to avoid selling the business one possibility is to divide it, where, assuming it can be restructured to allow a demerger, the next generation take over different parts that then develop independently. But companies should not take this route purely for family reasons – it must also make good business sense. Other options, particularly once the family business reaches the third generation and beyond, include appointing professional non-family managers. By this stage there can be dozens of family members with a stake in the firm, and introducing professional management often represents the only realistic solution to the succession issue. If the obstacles to family succession are temporary (for example when the successors have not yet acquired the experience to take over), a caretaker MD can be appointed to run the firm until the transition within the family eventually takes place.

The Next Generation's Perspective

The next generation in a family business have a unique opportunity to build a challenging and enriching career for themselves. But if they join for the wrong reasons – maybe searching for a safe haven – the decision is likely to be one they live to regret.

One of the few 'golden rules' for family businesses is that, before joining, the next generation should obtain outside career experience. This will help them to develop an objective view of their own talents and abilities, and will boost their credibility. Working for upstream, downstream or larger companies in the same sector provides practical industry experience, or joining someone else's family business for a few years (there are networks that organise this) helps generate a fund of ideas to bring back to the family firm.

The next generation should also beware of sibling rivalry. Where brothers or sisters work together in a business, some rivalry is normal, but efforts may be needed to prevent these feelings becoming a destructive force. This means managing sibling rivalry rather than being managed by it. A good idea is to agree

9 For more details, see (Bibko 2006) and (Asian Institute of Management 2006).

on a code of behaviour that recognises the welfare of the business is paramount, and that establishes procedures for resolving differences.

Preparing for Leadership

The next generation's role in the succession process centres on being thoroughly prepared, and doing everything possible to ensure the transition takes place smoothly. After joining, potential successors should fine-tune their training and development programme, encompassing all important aspects of the business. Because everyone working in, and doing business with the firm will know they are 'family', they will have to demonstrate not just their willingness to work hard but also an extra dimension of commitment to establish their own identity and gain the respect of employees.

'Emotional intelligence' – the ability to understand not just our own emotions but also the emotions of people around us – is an important skill in family businesses, and understanding this concept helps potential successors who are under scrutiny from the range of different stakeholders (Nicholson and Björnberg 2008). Mediation and communication skills also need to be cultivated, along with an awareness of the special symbolic value next generation leaders carry, and how this can be used to create value in the family company.

The ideal result will be that the next generation gradually takes on more and more responsibility so that assuming leadership, when the time comes, represents a natural progression. Juniors should talk to the senior generation concerning fears the latter may have about loss of authority, status and self-esteem. This appreciation of their perspective should help to reduce the potential for conflict and ease some of the main emotional problems of the succession process.

The next generation's role in the succession process also presents them with a golden opportunity to renew the strategic vision and values that underlie the family business. An early task should be to set about revitalising the history, culture and vision into a shared mission that can improve the firm's competitive advantage and give the family a fresh sense of cohesion and purpose.

Succession in Older Family Businesses

US family business professor John Ward was one of the first analysts to realise that the ownership structure in a family business is fundamental to understanding the real forces at work within it. Some 25 years ago he drew attention to the fact that, in broad terms, ownership of a family business tends to progress through a sequence, reflecting ageing and expansion of the owning family (Ward 1987). It evolves from the simple first-generation controlling-owner structure, through the sibling partnership stage (where ownership has been divided among a group of sons and daughters of the original family) to the complexities of the third-generation-and-beyond family business, called a cousin consortium.

The approach to succession from first to second generation is dictated largely by the character of the business founder, with all the issues made more complicated by the founder's dual role as parent and employer, as well as by their probably ambivalent attitudes concerning relinquishing control and coming to terms with the realities of age and mortality. Later generation transitions, in contrast, are much more likely to be governed by the size and nature of the company, and by market conditions.

Note, however, that ownership does not necessarily progress sequentially from one form to the next. For instance, not all owner-managed family firms are first generation businesses. There are examples of family businesses where the single-owner model is recycled, and the company is passed on to just one owner in the succeeding generation. Indeed, this happens a lot in the farming sector, where families do not want to split land holdings among siblings. Also, share buy-backs at family firms can lead to the re-establishment of either an owner-managed business or a sibling partnership, where one branch of the family buys out the others and takes control.

Second to Third Generation

Second to third generation transitions are often easier for the family to cope with. They will already have successfully come through the cross-over from first to second generation, almost certainly learning a lot from the experience. Also, members of the second generation have a number of factors working in their favour: the business they have inherited is up and running; the second generation will probably have received a better education than the first; they may well be more skilful when it comes to business management; and they should represent a source of fresh enthusiasm and vigour that can take the business forward.

Some successors, on the other hand, having grown up in a protected atmosphere of comfort and financial security, may not share their parents' dedication to the family business. They may have joined under a feeling of obligation, and their lack of motivation and commitment can lead to the firm's demise, often accompanied by deterioration in family relationships.

Second generation members face particular problems in relation to ownership of the business. Whereas the founder probably enjoyed both day-to-day control and 100 per cent ownership, his or her successors find themselves operating with a new leader, who they may or may not whole-heartedly support, and as co-owners, perhaps even with only a minority stake. And when they come to consider succession, they must face a similar type of problem to that with which the founder had to grapple, but on a much larger scale. There will usually be more succession candidates when the second generation comes to decide which of their children should take over the business (a situation often exacerbated by equal second generation voting power and a history of unresolved conflicts). Also, cousin relationships will have become an issue. Cousins will have, as one of their parents, an in-law who grew up outside the family, and he or she may hold radically different values as a result. This diversity means that in the transition from the second to the third generation, an important challenge involves developing effective communication, leadership and difference-management skills among cousins.

Third to Fourth Generation and Beyond

By the time the third generation is in place, there's a well-established business, and a widening circle of family members. An important characteristic of the third generation is its diversity. A range of in-laws is likely to have become involved as brothers and sisters have married people with widely differing values and perspectives, and they themselves have had children. The diversity can be such that it is difficult to believe that all the children come from the same family, and some will grow up loving the family business because they were taught to by their parents, while others will hate it because they feel trapped in it. At this stage it becomes vital to have some sort of escape mechanism in place enabling those who want to exit from the family business to do so.

Strategic issues in the transition from the third to the fourth generation, over and above those that affect every generation, frequently centre on a loss of direction and purpose. The drive, ambitions and objectives of the founder may all have become no more than an interesting piece of family history, or the original goals may have been overtaken by events in a changing world. At this point, either the company is actually sold or the vision must be recreated by members of the third and fourth generations revitalising the business by engaging the family's enthusiasm and commitment to its future.

More than just 'Changing the Guard'

We've seen how adopting a 'systems perspective' of family enterprises helps us cope with the idea of lots of things happening at the same time. I'll end the chapter with one final systems perspective concerning ownership succession. It is a dangerous mistake to regard succession as simply a question of transferring a tried and tested way of running the business from one generation to the next. Rather it's a *system* change; a transition to a different type of business structure with a different culture, different procedures and different ground-rules.

For example, if third generation cousins look to what their parents did in achieving success at the sibling partnership stage, and try to emulate it in the cousin consortium phase, they usually fail. It is easy to overlook the huge challenge that this implies. In effect one is asking owners to forget what they learned through decades of observation and example, despite the fact that they have masses of data proving that what it was they learned worked very well!

Also, these changes in system and culture do not take place overnight. Figure 12.3 shows how in most successions there is a transitional period during which the business is effectively 'between systems'. Depending on the spread of ages within generations, these periods of overlap sometimes last for a decade or more. So, in owner-manager to sibling partnership successions, as the transition progresses the business loses certain of its owner-managed characteristics (heroic culture, centralised decision making, and so on) and gains more sibling partnership characteristics (shared vision, effective teamwork, and so on). During the transition there is in effect a hybrid system, and this can be confusing for everybody. Around the mid-point in this succession transition there will typically be a group of siblings trying hard to become a team, but working under the owner-manager who not only finds it hard to understand teamwork, but may well see it as a sign of weakness.

A valuable aspect of this particular systems perspective on succession is that it helps to focus attention on the issue of inter-generational teamwork. It is the task of both generations on either side of the succession transition to come to terms with these confusing and often counter-intuitive ideas, and having such a pretext for collaboration is precisely the sort of inter-generational rallying point that helps families to negotiate transitions successfully.

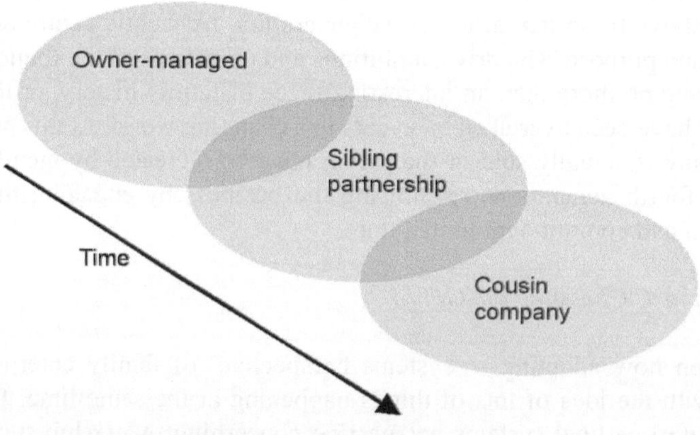

Figure 12.3 Transition Phases
Source: Adapted from Gersick, Davis, Marion McCollom and Lansberg (1997)

References

Aronoff, C.E. 2003. *Letting Go: Preparing Yourself to Relinquish Control of the Family Business*. Marietta: Family Enterprise Publishers.

Aronoff, C.E., McClure, S. and Ward, J.L. 2003. *Family Business Succession: The Final Test of Greatness*. 2nd edition. Marietta: Family Enterprise Publishers.

Asian Institute of Management 2006. GMR Group: Case Studies. Manila: The Asian Institute of Management.

Bibko, S. 2006. Good groundwork. *Families in Business*, September/October, 43–46.

Davidow, T. 2006. Reshaping – not retiring. *Families in Business*, September/October, 71–72.

Dorgan, S.J., Dowdy, J.J. and Rippin, T.M. 2006. Who should – and shouldn't – run the family business. *The McKinsey Quarterly*, 3, 13–15.

Gersick, K., Davis, J., McCollom-Hampton, M. and Lansberg, I. 1997. *Generation to Generation: Life Cycles of the Family Business*. Boston: Harvard Business School Press.

Hudson, F.M. 1999. *The Adult Years: Mastering the Art of Self-renewal*. San Francisco: Jossey Bass.

Institute for Family Business 2008. *The UK Family Business Sector Report*. London: Capital Economics Limited.

Lansberg, I.S. 1988. *The Succession Conspiracy: Mapping Resistance to Succession Planning in First Generation Family Firms*. Working Paper A70. New Haven: Yale School of Organization and Management.

Leach, P. 2011. *Family Businesses: The Essentials*. London: Profile Books.

Levinson, H. 1974. Don't choose your own successor. *Harvard Business Review*, November/December, 53–62.

Murray, B. 2005. From my hands to yours: retirement as self-renewal. *Families in Business*, November/December, 36–39.

Murray, B. 2009. Creating a new life after leaving the family business. *Campden FB*, Autumn, 44, 66–69.

Nicholson, N. and Björnberg, A. 2005. Shared leadership in family firms. In *Family Business Leadership Inquiry*. London: Institute of Family Business (UK), 18–20.

Nicholson, N. and Björnberg, A. 2008. The shape of things to come – emotional ownership and the next generation in the family firm. In *Family Values and Value Creation*, edited by J. Tàpies and J.L. Ward. London: Palgrave.

Okoroafo, S.C. and Koh, A.C. 2009. Family businesses' views on internationalization: do they differ by generation? *International Business Research*, 3(1), 22–28.

Schwass, J. 2005. An effective successor development strategy. In *Unconventional Wisdom: Counterintuitive Insights for Family Business Success*, edited by J. Ward. Chichester: John Wiley & Sons Limited, 97–115 at 109–110 and 114.

Tagiuri, R. and Davis, J. 1982. Bivalent attributes of the family firm. Reprinted (1996) in the 'Classics' section of *Family Business Review*, IX(2), Summer, 199–208.

The Economist 2004. Family business: passing on the crown. *The Economist Special Report*, 4 November 2004.

Ward, J. 1987. *Keeping the Family Business Healthy*. San Francisco: Jossey Bass.

Webb, P. 2005. *A Systemic Approach to Responsible Ownership*. Paper presented to the 4th National Forum Conference of the Institute for Family Business

(UK), Success Through Family Teamwork: Working Together Across the Generations, Birmingham, 19 May 2005.

Zalman, J. 2005. Expanding the vision. *Families in Business*, November/December, 70–71.

Chapter 13
From Generation to Generation: Drawing the Threads Together

Ian Whitehead, Matt Lobley and John R. Baker

Introduction

In the introduction to this book it was argued that, 'The long-predicted demise of the family farm has proved to be somewhat exaggerated'. In addition to a wide range of other key messages, the replication of the *FARMTRANSFERS* project across the developed world, along with the many other contributions to this book confirm that this is most certainly true. Indeed, the family firm remains the cornerstone of the farming industry, evidenced as 98 per cent of farms in the USA (Hoppe and Banker 2010), and, according to the Australian Bureau of Agricultural and Resource Economics, 2005 (Hamblin 2009), representing 94 per cent of farms in Australia. In a recent survey of farm businesses in the UK, 71 per cent of the respondent farm principals were there by virtue of succession (ADAS Consulting Ltd et al. 2004), of a similar order as the results of a survey in the south west England, mentioned in the introduction, where 81 per cent of respondents operated 'established family farms'. In Japan, the picture of farming families is unusual in comparison with other countries, with much decline in the numbers of farm households over the four decades to 2005. Despite this, the 2000 Japanese Census indicated that multiple generation families constitute 80 per cent of farms. Moreover, in the order of 8 million of the population are still recorded as living in farm households (Uchiyama and Whitehead, Chapter 4).

The key motivations for the continued existence of family farms are summarised by Miljkovic (2000: 544) as: the 'symbolic importance of the farm, usually increasing the subjective value of the farm far above its economic value; the farm as a place of residence as principal and in old age and the place where children' were raised; the existence of free human capital provided by the children as they grow up and the mutual interest and continuation of practices and 'care' that the holding will receive from the children. Price and Evans (2009), in their work on farming stress on family farms in Wales identify 'deep-seated geographical factors' as contributing to pressures to continue farming on that holding. Those cognisant of these drivers recognise the complexities of the mix, the idiosyncrasies of connections and combinations, the resilience of the 'organisation' and the importance of 'smooth' transference of the business to the next generation.

This chapter benefits from the rich expression of practice from countries as far afield as New Zealand, Australia, Canada and the US, all with Commonwealth influence in common, with contrast provided by replications from Europe and Japan. Analysis of this evidence first considers the continuing merits of the family farm, moving to a collective review of key contemporary challenges and the drivers. The common and differential practices of succession across countries form the focus of the next section, before a critical reflection on modes of intervention by government and others. This then provides the basis for the assessment of the role of family farming in future sustainable food and non-food supply across the globe and the potential for response from these businesses to a range of well recognised pressures and challenges.

Perpetuating the Strong – the Continuing Merits of Family Farming

Family farms vary in terms of size and other characteristics. For current purposes, it is the close link between family and business that is important – the 'familyness' of the business (see Chapter 1). The interest in the transfer of the family farm between generations by the authors of each of the chapters in this book is, perhaps, testament alone to the continuing importance of family farming across the developed world. Whilst economic and environmental pressures continue to create challenges for the farming industry, whether in Switzerland or Australia, family farms demonstrate resilience and fortitude, driven by factors that are less well known in other businesses.

Such strength, it is argued, arises partly as the result of familial ideology and what Pile (1990: 64) terms, 'the reproduction of the family', which in turn relies on succession. The thought of past generation(s) and future succession opportunities challenges the mind and, for many, drives the development of strategies in response to pressures and opportunities as they arise. Identifying additional connections of the family to the farm as a place, Pile goes on to note that 'the feeling of continuity is very strong in terms of family, farming and attachment to the locale' (page 64). Price and Evans (2009: 6) suggest that, 'personal hardship does little to undermine the sense of duty that becomes associated with sustaining patriarchal family farming', although clearly a source of stress.

The exposition of family *firms*, provided by Peter Leach in Chapter 12, emphasises the strengths of family enterprise in commerce *per se*. Family businesses have *a special status*, typified by 'long-term orientation, patience, common values, strong commitment and the idea of stewardship'. The nature of the family business is of long-term loyalty, care and the nurturing of family members, with consequential *special tensions* arising from the overlap between 'family and business domains'. In some instances, it is clear that such loyalty and commitment to family has weighed heavily for some, such as those in the Australian wool industry, where farming parents find themselves encouraging offspring to look elsewhere for a career. In other cases, the long term orientation and lifestyle features of family firms, described by Barclay et al. (Chapter 2) as 'the intimate connection between the farm as a place of work, career and family tradition', can stifle the necessary hand over of the business to the next generation. Emotional ties in the family context demand, therefore, equally *special treatment* in matters of succession, if the true benefits from family firms are to be achieved.

Changing Landscapes

The economic, social and environmental setting for farming businesses has changed dramatically in the last three decades, with consequential impacts on farming and family farming. Ward notes, 'We seem to be witnessing important changes in the ideology of 'family farming', reflected in a declining commitment to succession and driven by poor economic returns in farming and the changing role of agriculture in a 'post-productivist' countryside' (Ward 1996: 213).

So what is new and how has this influenced change in family farms? The dynamic nature of this period has, thus far, been characterised by the following:

- Variable farming incomes – the general decline in farm incomes seen in some countries (EU) has not been the case in others. For example, in the US, since the farm debt crisis of the mid 1980's, farm incomes have been steadily rising.
- Decline in the influence of farming and the public perception of farmers, with greater focus on the relationship between farming practice and the environment ('agriculture as the engine of destruction'). For example, the last two US Farm bills have included environmental requirements and there are 'more visible' provisions concerning conservation practices.
- Policy movement towards rural development, to encourage the diversification of resource use and the stimulation of employment opportunity.
- Further developments in technology, easing the need for labour on the land and enabling longer working lives (assisted, perhaps, by improvements in health).

Concurrent with these changes in the business environment, the pre-2008 fortunes of economies across the globe have encouraged further migration of incomers to rural areas, with related difficulties concerning the supply of affordable housing in rural areas and serious impacts, in some locales, on rural infrastructure, employment opportunity and schooling and other services (ADAS Consulting Ltd. et al. 2004). The buoyancy of sectors beyond farming, until recently, has also tempted, 'might have been' successors, away from the farming business. In addition, changes in society have arguably contributed to providing distractions away from the business. The incidence of infidelity in the modern world and the rise in divorce rates has increased concern over the depletion of resource availability on family farms as a result of marital break of the successor, referred to in Australia as the 'daughter-in-law problem' (Chapter 2).

For most career decisions the prudent decision maker will no doubt weigh the terms of the prospective position against expectations and alternatives. In farming some would argue that entry to farming has also thus been affected, 'succession has been strongly rooted in agriculture's 'productivist' rationale and throughout much of the post war period has been underpinned by steadily increasing asset values' (Ward 1996: 210). Many rapporteurs have confirmed the continued steady decline in farming incomes in real terms, although, again, this is not universally the case. Where the former is true, the appeal of farming to the young mind is potentially an issue, balanced against the attractions of alternatives, more obvious now than ever, through media and internet. Changing fortunes in the economy as a whole, since 2008, may have had an ameliorating effect. However, the oft-recognised 'security' of land-based occupations, linked to apparent comfort of family business has, perhaps, been elevated in significance in recent years. Indeed, it can be argued that in many instances notions of post-productivism have proved to be little more than 'academic fantasy' (Winter and Lobley 2009: 322) and that productivisim or 'new productivism' (Lobley and Winter 2009: 2) is back on the agenda and being enthusiastically embraced by farmers and their advisors. In the US, most of the programs that match beginning farmers with existing farmers, thereby providing a farm business successor, (albeit from outside the family), report that they have approximately 20 potential beginning farmers for every existing farmer, reflecting continued strength of interest.

Nevertheless, it would be naïve not to acknowledge the economic challenge posed by an international policy agenda that has promoted both liberalisation and the agri-environmental reform of agricultural policy regimes (see for instance, Potter 1998) and which, in some cases, has changed the relative significance of 'family' and 'business'. For some, such pressures may completely change the focus of attention: 'wider bourgeois business criteria are replacing traditional ideas of family continuity with ideas of farm business continuity, in which the family is involved but not central. In this way economic relations are becoming reconstructed.' (Pile 1990: 85). Such reconstructions may take different forms, at one end including suggestions of part-time off-farm work or the diversification of resources and at the other end, the necessity, rather than the desirability, as in the past, of fulltime work away from the farm for the meantime or, alternatively, for the foreseeable future. The *FARMTRANSFERS* project confirms the importance of off-farm work, where the size of the holding (Japan and Switzerland) or the economic conditions (affected by climate, as in Australia (Chapter 2) have necessitated this. In the US, the off-farm employment often provides the health insurance for the family. In a 2010 press release, the Kaiser Family Foundation reported that the average cost of family coverage health insurance was $13,770.

For many, this is where the *special characteristics* of farming and family farming, in particular, project the integration of life, work and tradition and the long term ideology of 'reproduction' and succession. For many there is clear appreciation of the impact of financial pressures on the business, and for some, clarity also of the likely returns from investment of the capital elsewhere. However, what is different for most family farms is that 'there is a value attached to farming 'beyond the economic' (Pile 1990: 147). This, it is argued maintains resistance to sale and alternative pathways, thus preserving continuity.

Although farm families have long supplemented household income with non-farming sources of income (Gasson 1988), a key feature of contemporary agricultural change is the search for alternative forms of income to substitute declining returns from agriculture. For the family farm, as well as others, this has required the reappraisal of resource use on most holdings – re-assessment of the appropriate use and combination of land, labour and capital for agricultural and non-agricultural purposes. Family farms have again responded to the call, 'farming communities are becoming socially and culturally integrated into new ways of thinking in the countryside of the 1990s (Ward 1996: 211).

Diversification in the use of farm resources has continued for many, as part of the succession plan for the future. For some prospective successors, this has added attraction by way of providing an add-on enterprise(s) to excite young, creative minds, holding them in the business. The contemporary imperative of 'getting closer to the market' with their products presents a considerable opportunity for the business, with prospective successors providing skills and drive, thus relieving, what some have characterised as 'conservative' farm principals from this challenge. For others, the suggestion of rationalising elements of truly agricultural enterprises may present a threat: a shift from the past and the recognised norm.

In his seminal publication, 'Country Life – a Social History of Rural England', Howard Newby (1988) refers to the 'key element of continuity' as stewardship and notes that 'the fertility of the land may be easily harmed by the short-sighted pursuit of immediate gains' (page 203). Concerns over the environmental impact of more intensive farming systems associated with productivist agriculture, have clearly tainted the reputation of farmers, contributing to their own sense of 'not being wanted' and being 'misunderstood (Lobley et al. 2005), and this may have influenced the attitudes of potential successors. Recognising the average age of family farm principals of 50–55 years old, most of the current farming fraternity have spent much of their time managing the business during the productivist era. Whilst additional challenges and opportunities raised by policy shifts are not solely the preserve of the prospective successor, the availability of opportunity to acquire the necessary knowledge and skills required by the 'new order' is perhaps greater for them, not least in the area of environmental sensitivity in farming.

In summary then, family farms have encountered a challenging period, with potential impacts on the place of family farming in the developed world. Indeed there is evidence from the contributors to this book that these and other pressures (for example drought in Australia) have most definitely reduced the level of certainty of continuity in some countries, although 'traditional norms and values continue to influence farm families'.

The Succession Process – Key Prerequisites

The identification and presence of a potential successor(s) is essential to the future of family farms. Without this there can be no succession. The second prerequisite is the release of control of the business by the incumbent business principal. This section reviews the features of these prerequisites through evidence available from both the *FARMTRANSFERS* project and other contributors to this book. Common threads will be drawn together, and variance from the norm highlighted within the context of changes to the setting for family farms over the last three decades.

The Successor

An obvious first place to start is the identification and presence of a successor, which is dependent on a number of factors, key amongst which are the presence of children in the family (or substitute close relatives) and the desire or motivation of those children to succeed in and with the business. The presence of a successor, alone, has been noted as having an impact on the decision making in the business (*the succession effect*), with a close clear correlation between this and the propensity for the principal to have plans for semi-retirement (Potter and Lobley 1996; Lobley, Baker and Whitehead 2010). The availability of adequate capital has been identified as less of an issue in the family farm scenario, especially when compared with the new entrant entering farming outside the family succession route. The availability of skills is also less of an issue (ADAS Consulting Ltd et al. 2004).

In all of the countries considered in this book the centuries old custom of primogeniture (the inheritance of the entire estate of the parents by the first born and in most cases the first male born), still has a strong influence over succession on family farms. As Price and Conn argue in Chapter 6 (Northern Ireland), the desire to keep the family name on the land though patrilineal succession remains a powerful driver. There is evidence, however, that farmers also value careful consideration of all children in the family, with preference given to those who are most interested in taking the business on (Barclay et al. Chapter 2). Further, anecdotal evidence from the US suggests that the number of prospective female successors attending business succession seminars has been rising.

As Leach points out (Chapter 12), 'the next generation in a family business have a unique opportunity to build a challenging and enriching career for themselves'. This is another of the strengths of family farms – the sequential nurturing and mentoring of the replacement principal for the business, before and after hand-over. After hand-over, with the former principal in 'semi-retirement', labour (*pro gratis*) is often provided by the parents and in some cases, the parents may also take on a mortgage to allow the successor to restructure the holding (Switzerland, Chapter 5) The motives of successors and their keenness to have a deeper involvement in the business will vary and be shaped by a number of influences. Amongst the most important is the degree of enthusiasm and guidance of their parents and principals in the business, articulated by Pile (1990: 64), where, 'previous generations provide the background for the making of a farmer. This can be either a coercive force or a desire to follow in the family, usually the father's footsteps'.

Perceptions of past successes or failures and future prospects link directly to the state of the industry. Our contributors are almost unanimous in the view that this is not only affecting demand from farming children, but also steadily reducing the availability of opportunity through farm sales, summed up by this quote, 'today, leaving the farm to your children is a form of child abuse' (Kaine et al. 1997; see Chapter 2). The impact of this is not always absolute in terms of succession. Sons and daughters are encouraged to better equip themselves through education and off-farm working, with the potential to return full time to the farm in the future, sooner (as returns from farming improve) or later (Australia, Chapter 2). In Japan (Uchiyama and Whitehead, Chapter 4), evidence suggests that returns from farming are so low as make most holdings dependent on off-farm income and, in some cases, to relegate motives for succession to the prospect of capitalisation of the asset through sale for development. In the context of land abandonment in this country, the family farm has kept the land in production (albeit encouraged by particular fiscal arrangements), another clear strength of the family farm set up. By way of contrast, in the US high commodity prices and low capital gains taxes have driven investors to purchase farm land, thus inflating land prices. Whilst this can enhance capital sales for the retiree, for the successor this may constrain expansion through land purchase.

As well as the likely economic prospects of the family farm, the potential successor does not live in a social vacuum and is variously affected by the changing status of farming as part of wider socio-cultural change (Newby 1978). It is certainly the case that in England, the public perception and 'standing' of farming over the last three decades or so has shifted from an industry of successful post-war domestic food provision to one seen as 'the engine of destruction of the environment', with questions raised concerning animal welfare and the distorting effects of subsidisation through the EU Common Agricultural Practice (Potter 1998). As Marion Shoard memorably asserted, 'The landowners' carefully nurtured base of public acceptance has been kicked from under them as the media have woken up to the damage being done to the countryside' (Shoard 1987: 549). This is reflected by Rossier in Switzerland (Chapter 5) and has, until recently, most definitely contributed to the decline in demand for agricultural education in England during this period, with subsequent depression in the interest in farming as a career. In stark contrast, in countries with a stronger agrarian ideology such as Australia, there is still a 'strong belief in the family farm as the appropriate production unit' Barclay et al. (Chapter 2).

Perceptions are clearly balanced by the lifestyle qualities often mentioned by those working the land and evidenced by Rossier in Switzerland (Chapter 5), 'activities such as outdoor work and working with animals' and 'the family structure characteristics of the work environment'. The *FARMTRANSFERS* replication in

Switzerland illuminated a further influencing factor in the succession decision, the 'duty of care' for the parents or in-laws on the farm (perhaps less strong in modern times), and the accommodation of the retiree on the farm post-handover. This is not mentioned by other contributors but if this is a common phenomenon, it may represent a further strength of family farming, perhaps hitherto unrecognised, with very definite tangible benefits to society.

Two principal routes for potential successors have been identified, namely: (1) the *direct route,* where successors go directly into farming after they leave school, and (2) the *diversion route,* where successors are employed in an off-farm job after leaving school and then return to the home farm operation at a later date. This is sometimes referred to as a professional detour (Gasson and Errington 1993; Uchiyama et al. 2008). The succession route followed is likely to be influenced by a number of factors, including the availability of alternative employment and cultural norms regarding the value of nonfarm work. Uchiyama and colleagues found that farm size is a predictor of succession route (Uchiyama et al. 2008). In addition, successors who are from smaller farming operations are more likely to be employed off the farm (that is Japan and Switzerland). In contrast, in England 62 per cent of successors worked on the farm full time (Lobley, Baker and Whitehead 2010). Successors in the U.S. (and Canada) are more likely to take a professional detour route – a non-farm job right out of school, before returning to the farm operation.

In summary, it is evident that despite changes in agricultural support regimes, challenging economic environments and socio-cultural changes that may erode the social standing of farming, in many cases there is still strong demand from family members to succeed to the family farming business. It would be true to say that the structure of farming has changed during this period, involving the further polarisation in holding size (the so-called disappearing middle) and a good degree of diversification of activities on many holdings. Polarisation has provided opportunities for some, where land is released for expansion of the farm, and challenges for others, where the farm is small or has been reduced in size. In latter cases the successor is more likely to be required to combine work on the farm with income generation off the farm until or beyond the retirement of the principal. It is clear from this analysis that the traditional values attaching to farming as a family and retention of continuity provides a strong position from which to develop responses to the range of challenges facing agriculture.

The Release of Control of the Business – 'No One Can do it as well as I Can'

In simple terms, the succession process is one of learning, leading and then leaving the business – 'leaving' the business for some other endeavour or activity, and 'leaving' the business for the successor to take on. Thus, the second prerequisite for successful succession is the willingness and ability of the incumbent to leave. Ultimately, this requires a mental withdrawal from the business as much as a physical withdrawal. Foremost in any succession process is the presence of a

Succession Plan, referred to in the USA as the 'Estate Plan'. For this to occur there must be some recognition or intention, ultimately, to retire, and then the formalisation of that in discussion and writing. The various contributors to this volume are unanimous in their submissions. In the developed world, covered by this project at least, the keenness to 'let go' is not common amongst the principals of family farms, although there is some variation between countries.

As Baker and Epley point out (2009) the decision to retire from the business may involve one of three intentions or actions (including the intention to never retire):

- To retire: you will provide neither managerial control or labour to the farm.
- To semi-retire: you will provide some managerial control and /or labour to the farm.
- To never retire: you will maintain full managerial control and provide some labour to the farm.

Evidence from *FARMTRANSFERS* surveys (Figure 13.1) indicates that farmers in Iowa (31 per cent), (and other US states such as Virginia (42 per cent), and North Carolina (47 per cent)) are more likely to remain employed on the farm operation, are less likely to semi-retire from farming, and will never retire. These figures are comparable with those for Japan, where 44.5 per cent of respondents plan never to retire. We have seen from Chapter 4, how the structure of farming (small sized units) and the high degree of interconnection between farm and off-farm work contribute to this.

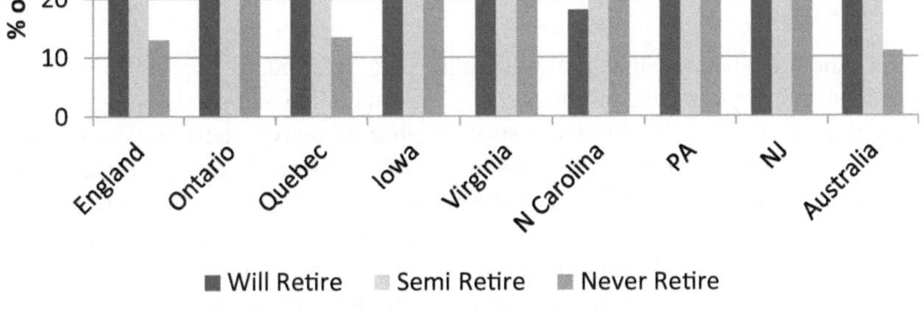

Figure 13.1 The retirement plans of family farmers

Farmers in Australia[1] (89 per cent) and England (86 per cent), and, Ontario (76 per cent), and Quebec (87 per cent) are more likely to plan to semi-retirement or fully retirement from farming. For these countries, semi-retirement is by far the preferred option for farmers, according to Foskey, known as *retirement in farming* (2002). Kimhi and Bollman (1999) suggest, for Canada, that the rate of exit is lower for larger farms and principals operating mixed livestock farms. The presence of a successor might make semi-retirement a realistic option for farmers who may otherwise face a choice of continuing to work full-time or completely retiring. For that successor, semi-retirement can present a number of issues. On the positive side there is the security that the knowledge and experience of the semi-retiring principal will still be available to the business. This is, however, potentially discounted to some degree by the realization that the control of the business will not be absolute, with the risk of unwanted 'interference'. This situation clearly requires understanding, trust, clear articulation and agreement. As for Japan, the high incidence of successor involvement in off-farm work most probably contributes to Foskey's, '*retirement to farming*' category of retirement (Foskey 2002) as well as the attraction of a higher earning career, with the prospects of returning to familiar farming roots later in life, as suggested by Wilkinson (Chapter 3) in the wool growers in Australia – the 'second career' farmer.

At the other end of the spectrum, away from the USA and Japan in terms of 'father move over', is Switzerland, where most farmers plan to *retirement from farming*, largely the result of and in accordance with Government policy, where the payment of essential direct payments ceases at the age of 65 and the quaintly titled 'old-age and survivors insurance' (AHV) is receivable. However, as in the replication in Japan, retired farmers seldom cease work in retirement. Working in retirement, means genuine quality of life, especially, as Rossier (Chapter 5) records, 'if the work is done voluntarily!' The Swiss policy to directly encourage retirement through manipulation of subsidy payment will be returned to later in this chapter.

The ability to finance retirement is likely to be one of a number of factors influencing retirement plans. Figure 13.2 presents comparative data on anticipated sources of retirement income and illustrates a number of significant differences

1 It should be noted that the Australian data referred to here is from a smaller sample than that reported on in Chapter 2. The Australian sample is comprised of two surveys; a main survey (N=789) and a smaller close of survey questionnaire sent to non-respondents (N=391), giving a total sample of 1180. Chapter 2 uses the larger sample of 1180. Comparisons conducted between the responses to questions that were common to both questionnaires were used to develop a weighting procedure to correct for the over- and under-representation of particular types of respondents in the main survey data. The weighting procedure was primarily employed within the analyses where the findings were generalised to the farming population. However, where the focus was on relationships between variables, unweighted data were used to avoid compromising the validity of tests of the statistical significance of these relationships.

between *FARMTRANSFERS* replications. The two Canadian replications are notable for the significance of the sale of farm land or other farm assets in order to fund retirement, reflecting the requirement to release capital for planned semi-retirement. Farmers in France, on the other hand, gain the largest proportion (48 per cent) of their retirement income from social security payments (Lobley, Baker and Whitehead 2010), while farmers in England tend to gain a significant proportion of their retirement income from private pension provision. In Australia, farmers 'tend to move to town' (Barclay et al. Chapter 2), allowing for continued involvement on the farm. There is no clear trend in the source of income relied upon after retirement (semi-retirement in most cases), with farmers dependent upon all sources, no doubt in different combination.

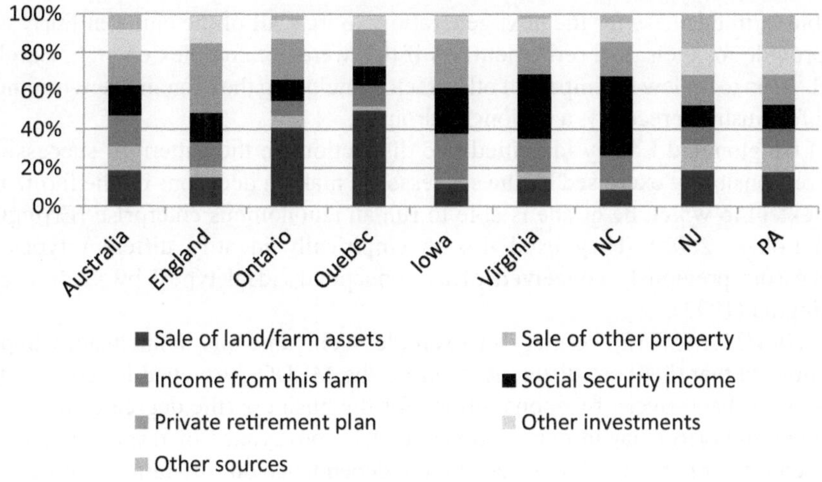

Figure 13.2 Anticipated sources of retirement income

Even where farmers intend to retire, the process of succession leading to the ultimate hand-over may be delayed, thus influencing the immediate or even the ultimate strength of the business in new hands. Pile (1990: 70) highlights some of the potential difficulties here, 'patriarchal beliefs in the *restricted* abilities of (women and) children' delay' or 'slow' the decision to pass on patriarchal authority which means that the handover of control over the business is usually seen as being gradual by the farmer.'

Based on experiences away from farming, Leach (Chapter 12) advocates a 'systems perspective' to assist with the complexity of the succession process. It is argued that the phrase 'handing over the reins' misinterprets the process which, instead, involves a change of leader adopting the principles and practices *of the day*, in order to steer the business through the next series of challenges, taking advantage of opportunities on the way. The wisdom of this is that it points away from merely handing over the reins of an existing system and involves the reshaping of the attachment of both retiree and successor to the business, recognizing that severance of contact is in most cases unlikely as the business 'is part of the fabric of the family'.

The Succession Process – Smoothing the Way

'Succession is a process, not an event and it should be carefully managed and take place over time' and 'should involve a well-planned partnership with the next generation' (Leach, Chapter 12). Academics and practitioners in this field have identified the complexity of succession planning. In farming businesses, Barclay, Reeve and Foskey (Chapter 2) point to the three main objectives: to maintain a viable farm business for the next generation, to treat all of the children fairly and to provide for their own retirement. As if this were not complex enough, Barclay et al. go on to review the impact of other factors including the economic environment and for Australia recently, a prolonged drought.

Errington and Lobley identified two distinctions in the pattern of succession: the responsibility exercised by the successor in making decisions on the farm, and the extent to which he or she is able to run an autonomous enterprise (Errington and Lobley 2002). They used this to empirically identify different types of successors previously conceived of as conceptual 'ideal types' by Gasson and Errington (1993).

FARMTRANSFERS results (for example, Uchiyama and Whitehead, Chapter 4) indicate that there are differences across the MEDCs surveyed in terms of the process of hand-over of responsibilities for the business (the degree of decision-making authority). Japan demonstrates a 'flat' conveyance of responsibility, the consequence of small holding size and the dependence on 'off-farm' income.

Successor experiences in England demonstrate what is referred to as 'Farmer's Boy' characteristics relating to hand-over. The successor who has worked with their parents for a long time but has been given few managerial responsibilities was termed the 'farmer's boy' by Gasson and Errington (1993). They argued that the farmers boy is a typical problem in farm succession in that the farmers boy has little opportunity to develop the managerial skills needed to operate the family business and is essentially a hired worker, kept in place by the promise that the eventual reward will be ownership of the family farm. Thus, in the case of the farmer's boy there is a very definite gradient of hand-over, from the conveyance of and maybe some responsibility for technical skills, at one end of the spectrum of the managerial function in the farm business, to much lower delegation of responsibility for the financial management of the business at the other end of the spectrum. This is a trend across all surveyed countries, confirming the patriarchal way of life in many family farms, although more profoundly demonstrated in England.

The second major category is the *Separate Enterprise,* where the home farm operation is large enough to support a separate enterprise run by the successor. This category allows the successor to develop managerial skills and also allows for some financial autonomy (Gasson and Errington 1993). The third category of succession route is the *Stand-by Holding,* in which the successor is set up on a separate farm in order to develop his or her farming skills. Although the successor might share machinery or labour at some point, he or she still remains independent of the farmer. Compared with other countries, English and Canadian successors are more likely to run a separate enterprise to develop farming skills necessary for farm operation (Errington 1998). Few US successors run a 'stand by farm'.

The identification of a successor is not always an easy task, as discussed above, with a wide range of factors, including the contemporary distractions of a more lucrative existence. Evidence from the *FARMTRANSFERS* project suggests that the identification of a successor varies across countries, from 52 per cent in England to 25–30 per cent for the US states in the project. A clear association is recognized between this and a path of semi-retirement from farming, in that those farmers who have identified a successor are more likely to experience some form of semi-retirement. This trend occurs regardless of nationality.

A good degree of attention has been given in some countries to determine approaches of smoothing the way, or ensuring that Goeller's three legged stool (see Chapter 9) is 'stable', whilst recognising the sensitivities whilst achieving the objectives. With predominance of such businesses amongst More Economically Developed Countries (MEDCs) 'how these families plan and manage retirement, succession and inheritance is, therefore, a concern for the whole agricultural industry'. The handover process has been conceptualised by the Iowa State University and the University of Nebraska (Goeller, Chapter 9) as involving four Phases of Transition: the Testing phase, the Commitment phase, the Established phase and the Withdrawal phase. The adoption of a 'business' approach is advised, endeavouring to develop the relationship from a purely paternal one (in many cases) and the model is offered as an appropriate approach.

Leach (Chapter 12), however, confirms that in family businesses generally, there is often 'a natural aversion to planning' with reasons often 'rationalisations designed to avoid deep-rooted anxieties and fears'. The apparent aversion to planning is confirmed by Ilbery et al. (Chapter 7) noting minimal uptake to the 'retirement and succession planning' initiative in Cornwall, England as part of the Fresh-Start initiative, established in 2004.

Understanding all of these issues is fundamental to smoothing the way for all parties.

Distinctive Features – a Typology of Family Farms

Thus far, in reviewing the characteristics and motivations of the parties in the succession process on the family farm, a number of the key differences have arisen from analysis of the results of replications across MEDCs in the

FARMTRANSFERS project. Socio-economic and cultural variations, past and present, have shaped the mix of values and beliefs as well as expectations of those in different countries. When combined with the complexities and dynamics of the business environment over last four decades. It is perhaps not surprising that a degree of commonality is exhibited by family farms across the developed world (Figure 13.3) and a number of these have been highlighted in this and previous chapter. However, reference to 'family farms' *per se*, perhaps as an indicator of certain types of farming business, is over-simplistic (see Chapter 1). It is clear from analysis of evidence from previous research, as well as that provided in the *FARMTRANSFERS* project that, as a result of continuing pressures, economic and climatic, the nature and practices of family farms of the contemporary developed world should perhaps be more usefully divided into three categories, the *lifestyle family farm*, the *multifunctional family farm* and the *agribusiness family farm*. This typology provides greater clarity in terms of different characteristics and motivations. In turn, such differences impact on key aspects of the succession process including the presence, motivations and preparation for succession of potential successors, the likely expectations and actions of the principal in the business and the nature of the succession process.

The typology has been constructed with the understanding that its purpose is to differentiate between family farms. In terms of the language used in Chapter 1, it is an *ideal type* approach rather than an attempt to develop an operational definition. It is recognised that, even within one country, there will be exceptions to what might be considered a particular 'type'. We are, therefore, endeavouring to highlight key differences in a general manner, whilst attempting to observe differences in succession characteristics between types. It follows also that each farm type differs in terms of the nature of any support required to ensure the smooth and timely transfer of the business.

By way of explanation, a pen portrait of each of the three types has been prepared. First, then, is the Lifestyle/Residential Family Farm (LFF), perhaps typical of many of the small Japanese farms or the mountain farms of the Switzerland. There is a tendency here for these farms to be small, by international comparison, and totally owner-occupied. The objectives of occupation relate primarily to residence and income from farming is subsidized by off farm work. The degree of intensity of agricultural production may vary considerably say between the Swiss mountains and the paddy fields of Japan, although with a tendency for less time to be spent on farming, exacerbated where semi-retirement is the case. Primarily as a result of off farm income, the economic vulnerability of such farms is low.

Broadly speaking, the succession route on LFFs is usually via a 'professional detour'. Successors tend to be involved in other off-farm work, which is apparently increasingly encouraged by parents, before returning to the farm, sometimes late in the life cycle with the intention of 'retirement to farming'. Again due to the size of the business, the principal in the business is less likely to want to retire and where semi-retirement is taken, this typically occurs later than in the other types of business. As a result of both these considerations, the succession ladder is

	FAMILY FARMS		
	Lifestyle / Residential	**Multifunctional**	**Agribusiness**
examples	Japan, Switzerland	England and Wales and US	dairy units and sheep stations NZ; prairie and ranchlands Australia and US
scale	very small/ small	medium scale	large / very large
land tenure	owner/occupation 100%	mixed tenure – agric. and non-agric.	mixed tenure - all agricultural
ownership objectives	property as residence; land as investment	diversify use resources to add value and revenue	priority on agricultural production; some non-residential
variety of resources	low	medium	high
dependence on agricultural income	low / much off farm income	range of sources/some off farm	high / total
intensity of agricultural production	high	varied	high
economic vulnerability	low	medium	high
time burden of holding	small	considerable	high / total
COMMON VALUES	**continuity of property ownership and business; familial ideology; patriarchal; value the family working environment; work / home integration; deep attachment to place;**		
COMMON STRENGTHS	**long term; knowledge of the land and appropriate practices; hard-working; greater propensity for community involvement;**		
succession route	professional detour / off farm work	professional detour / farmers boy	JVenture, separate enterprise or standby holding
ID of successor	later	varied	earlier
responsibility ladder	less well developed (flat)	varied	good clear ladder
exit rates	low	low	higher
retirement plans	high non retirement and later	semi-retirement	semi-retirement or earlier full retirement
attitude to estate planning and farm assets	assets seen as similar to other heirlooms - the business is of secondary import	assets seen as a legacy for passing down, probably in equal shares	asset ownership and management responsibility may be separate and succession viewed separately
retirement type	retirement to farming	retirement in farming	retirement from farming
succession effect	low	high	medium
successor effect	low	high	medium
retirement effect	high	medium and varied	low

Figure 13.3 Typology of family farms

much less well developed, presenting in some cases the proverbial 'steep learning curve' to the successor. These are the key features of the Lifestyle FF. Most of the common values and strengths detailed in the typology are shared with the other two types, with the exception of circumstances where the property is seen as an asset for sale. The differentiation of this group will be seen as useful, as the pressures, challenges and opportunities are quite different from the other two types. This has particular implications concerning issues of support for the FF type *per se* or for facilitating the succession process (discussed later).

In the middle ground, in many ways, is Mulitfunctional Family Farms (MFF). These are the farms which have diversified, many in the past three decades, in response to the market and policy direction. In many cases these farmers are farming 'middle ground' in terms of land quality and the potential to improve production is constrained by this or some other factor or combination of factors. The range of resources is greater, in many ways due to scale, and much effort has been applied to diversify the use of these resources such as by adding value through processing the primary product on the holding or providing one or more services to the public. Thus on these holdings there is an intimate (although sometimes less than complementary) combination of agricultural and non-agricultural enterprise. In order to keep this mix of enterprises performing, family members spend much of their time working in the business.

In terms of succession, again the characteristics of MFFs feed through to some common trends. First, exit rates and the expectation of full retirement of the principal are low. As a result, potential successors who may have been identified early on, are encouraged to achieve in education, in order to provide alternatives for employment and the professional detour route. In other cases, potential successors are employed in the business from an early stage and the combination of factors and the patriarchal spirit of familial succession can be associated with the 'Farmer's Boy' route. Here, although the succession ladder may be clear, the opportunity to step up to the higher rungs (of financial management) is left until the transfer by semi-retirement or the death of the principal. In this type of business, if the farmer's boy route has been avoided, the potential for the 'succession effect' and the 'successor effect' is high. Potential successors who have been educated to Secondary or Diploma/Degree levels, and maybe work on other farms, bring back to the business a range of ideas and skills to take the farm forward. This will better equip the business in terms of adaptation in the future. The successor may be the one to take on the diversified enterprise or appreciate the opportunity for other multifunctional activity to generate income.

Finally, firmly wedded to the principal objective of 'primary' production, and in many cases having positioned themselves over the last three decades to this end, is the Agribusiness Family Farms (AFF). Here, scale is generally greater, either a result of growth by successful previous generations or aided more recently by the gathering together of a 'mixed bundle of rights' – land licensed, leased or shared with the owner, in addition to an owner-occupied core. For many, dependence on agricultural income is absolute, intensity of production is high and sensitivity to changes in production and economic fluctuations is greatest. Although living on the holding is usual, non-residence is possible.

The principal and practices of patriarchy are perhaps strongest here and the masculinity of farming perceived as greatest on large dairy units and sheep stations and cattle ranches in New Zealand and Australia, respectively, along with the prairielands and ranchlands of Canada and the US. These farms are likely to represent the primary productive core of the future, vying for position with the corporate operations, in the delivery of food and non-food primary products.

As for succession, historically, primogeniture has been the norm, although 'business is now business' and other possibilities are conceivable. The survival of the business, rather than the family line, has perhaps taken over as number one priority. The potential successor will have been identified early and will have spent most of their lives in the business, with time out for training to diploma or degree level, most likely, in an agricultural subject. Although as everywhere, there is the possibility for the Farmer's Boy route to be followed, the size and scope of AFFs typically provide a clearer succession ladder with greater opportunities for delegation of decision making and progression. Early partnership or other joint venture arrangements are evident, with the business as a whole or with separate enterprises or sections of the holding, within it. Unusually for family farms, the likelihood of full retirement is greatest here, with perhaps less emotional connection to the 'place'. Once again, the identification of such different characteristics in this type of family farm will demand different approaches to policy and intervention for the farming and succession processes typically found here.

Tradition, Custom and Policy Supporting Family Farm Succession

Reference has already been made in the Introduction to the focus of attention of Government on family farms. In 1983, Harrison, highlighted policies across the then European Community in support of small farms concluding that 'it might be surprising … that little concrete evidence is to hand to justify such measures. Instead, it is evident that the reasons such policies have been pursued is that they command (or at any rate are hitherto commanded) widespread support. Partly, this is because large numbers of people have benefited from them directly. Partly, it is because such policies have a profound, above historic, nationalistic, almost jingoistic appeal.' (Harrison 1996: 135). Ten years later, in 1993, Harrison has become more positive, suggesting that, 'there may be good reason to think that there will be a greater need for family farms in the future than the 'market', as it currently operates, is taking into account' and that, 'there may be good grounds for believing that family farming bestows valuable benefits on the community at large, for which its owners and operators receive no reward. In modern parlance, this suggests that family farms 'give rise to positive externalities'. Justification is offered for the existence of such externalities including:

1. Small family farms 'absorb economic misfortune' while larger ones pass failures on to other members of society.
2. The more smaller farms there are the larger will be the rural population and 'the more efficient the use of social capital and rural infrastructure'.
3. Less likely to produce a 'monocultural pattern' of farming and more likely 'to produce a varied and aesthetically pleasing landscape with less environmental pollution'.
4. Likely to stimulate 'initiative, independence and innovation and contribute to the wider sharing of property ownership' (Harrison 1996: 136).

Evidence from the *FARMTRANSFERS* project suggests that where these externalities exist, they may have been weakened by the state of farming over the last three decades. Recognising the differences in human and physical geography across the *FARMTRANSFERS* project, this section is not an exhaustive review of intervention but instead seeks to provide an illustration of custom and policy supporting family farms and the succession process across the developed world. A variety of different approaches are evidenced, where government or others deem them to be appropriate, including financial assistance to smooth the transfer and help with restructuring of the business, customs dictating a range of inheritance paths and approaches through fiscal policy to help with capital exchanges from principal to successor. In terms of other practicalities, the setting for land tenure arrangements may be very important, as is the provision of matching services and education programmes to bring parties together and to effect well informed decisions before and during the succession process.

Financial Support for Farming

Any discussion of policy intervention must be prefaced by an acknowledgement that across many but not all, MEDCs (for example USA, EU and Japan), support for farming is a long established feature of government policy. The purpose of this section is not to consider government support for agriculture in a broad sense but to identify policy interventions that are directly or closely allied to the process of succession and retirement. A recent comparative study investigated financial distress surrounding farm family transfers in a number of EU countries (van Bommel et al. 2004) and found that the fiscal and regulator costs incurred in the UK (strongly family farm based) were low compared with other member states. In addition, inheritance and succession practices in the UK were relatively favourable to successors, as parents and siblings had no formal right to compensation when farm ownership was transferred. Recent research in the UK (ADAS Consulting Ltd. et al. 2004), considered the potential contribution of a range of difficulties experienced by potential new entrants. The study concluded that, 'it is not clear that there is any economic case on efficiency or distributional grounds for public intervention to influence entry or exit' (page 15). In itself, this statement reflects the value attached to intrafamilial farm succession, the mainstay of UK new entrant opportunity. In addition, in difficult economic circumstances, the externalities discussed earlier must continue to contribute to the position of family farms with regard to support, *per se*.

Within the EU member states, support for early retirement and young farmers has been available for some time, with the aim of accelerating the restructuring of the farming sector and reducing the rate of ageing of the farming community. Evaluations of these schemes (Caskie et al. 2003; Naylor 1982; Bika 2007) have, however, suggested that there is little quantitative evidence of their effectiveness: 'At the EU-12 level there was virtually no change in farmer age structure' with the figures for different Member States showing 'no positive correlation between expenditure [per holding] and change in farmer age structure' (AgraCEAS Consultants 2003). Notwithstanding this, a sequence of revisions of these schemes has ensured continued support across the EU (that is Greece, Ireland, Norway, Finland and France, although excluding the UK) – (see Chapter 6).

Inheritance – Custom and Law

The complexity of custom and law concerning the inheritance of rights to the deceased's estate is clearly articulated by Bika (2007) as varying according to class, geographical region and, in some cases, with time. For some cultures, the custom has been for inheritance of the whole by one beneficiary, male or female only, or either. In other cases, a more egalitarian approach is the custom, with the share of the estate equally between children. France adopts the middle ground, with farm successors paying compensatory amounts to coheirs, thus retaining the unity of the holding.

Primogeniture, the passing on of the estate of the deceased to the first born in the family, is predominantly an Anglo-Saxon approach to succession and therefore common in those countries touched by colonisation of previous centuries. This, combined with the partriarchy of farming in MEDCs, has secured the passage of the first born son in family farms for centuries and is clearly still strong in a number of the countries involved in the *FARMTRANSFERS* project (UK, US, Canada and Australia). The greatest strength arising from this tradition is the preservation of the holding as a whole, subject, in some cases to the release of capital, evident in the Canadian provinces, for example. A degree of decline in this practice is, however, noted, where the first born is attracted, perhaps by other more lucrative careers elsewhere or the parents are concerned over the possibility of handing over a 'poisoned chalice'. This has been the case in the UK (ADAS Consulting Ltd. et al. 2004) and is referred to by Barclay et al. (Chapter 2) and Wilkinson (Chapter 3) in Australia and on the smaller holdings in Switzerland (Rossier, Chapter 5).

In other cases, children who are the most interested in farming are favoured or the parents are simply driven by equity to provide an equal share of the inheritance to the children. For this to occur, the farm may be leased by other members to the family member who is most interested in the continuation of the business on the land. The latter can be at the heart of discussions, where one generation or more have built up the holding or developed first class breeding lines in livestock, for example. In other cases, the land, and farming assets are simply sold to provide the capital for sharing amongst offspring.

An interesting variance on the question of farm inheritance is evident in Switzerland (see Chapter 5). Here, direct payments are receivable until 65 (full handover). Also in an endeavour to promote family farm succession, the value of the holding on conveyance from principal to successor is prescribed by law as the capitalized earning capacity of the holding. As suggested by Rossier, this can leave the principal with a problem, whereby the asset released is less than the market value of the property. Further pressure can come at this time, as the state pension scheme exempts farmers (as self-employed) from the requirement to subscribe to an additional occupational pension plan and itself does not cover the costs of living. In many case therefore the principal is dependent on income and housing from the farm, not a conducive situation for maximum succession effect!

Perhaps more of an encouragement, in Japan, the Farmer's Pension Scheme, has provided additional Government contributions to the scheme for young farmers, payable on transfer of the farm business to successors at 65 years of age.

Fiscal Incentives

A range of fiscal provisions are possible to assist with succession. In the UK, taxation on the transfer of the estate of the deceased, once referred to as 'death duty', has long been seen as a major factor in the break-up of the large landed estates of the nineteenth century. In more recent times, this has been replaced with a much less draconian approach to Inheritance Tax. Here, providing the estate is transferred to the beneficiary(ies) and the benefactor survives this transfer by seven years or more, the tax is totally avoidable. Whilst not strictly targeted at farming or succession, the benefit to succession is clear – the estate of the farming principal is not broken up in order to pay tax.

Perhaps considered as one step further than this, in Japan, in an endeavour to keep small holdings from becoming smaller, the Government exempts farmland transfer from inheritance tax where the transfer is to a sole successor. With slightly different objectives, the US (Goeller, Chapter 9) refers to a number of states, including Nebraska, that have legislated for Beginning Farmer tax credit programs, encouraging landlords to let land to new entrants.

Land Tenure Arrangements

The flexibility of land tenure arrangements (relating to agricultural and non-agricultural property) can be a major contribution to the succession process, in providing a means of keeping the occupation of the land as a whole, for example, where the freehold for the property has been transferred in equal shares. In addition, flexible legal provision may also offer the opportunity for growth of the business, or its diversification, allowing the business to support more than one family through the succession stages. By way of example, such flexibility was provided in England and Wales in 1995, with fundamental review of the agricultural tenancy legislation in these countries. This moved the nature of the agricultural tenancy from a prescriptive tenant-orientated provision to a more flexible tenancy, in the hope of freeing up more land to provide greater opportunity for new entrants. In effect, the new tenancies have provided opportunities, not for new entrants to farming (that is starters), but for existing farmers (and their successors) to enlarge their businesses as mixed tenure holdings, with a clearer facility to incorporate diversified enterprises on tenanted land (Whitehead et al. 2002).

Matching Services and Succession Education Programmes

In England, successive governments have declined opportunities to introduce new entrants' schemes for agriculture. However, following the 2002 Curry Report on the Future of Farming an industry-led initiative, Fresh Start, was established. Although DEFRA have been the main sponsors of the project, additional support has come from the banking sector and National Farmers' Union and a national steering group includes representation from a range of agricultural sector organisations. Although (Ilbery et al. Chapter 7) describe unique interpretation of the Fresh Start initiative in Cornwall as 'largely unsuccessful' in terms of facilitating non-familial successional entry to farming, the national scheme, which is more focused around a series of some 20–25 Fresh Start Academies and less concerned with facilitating farm exit, is estimated to have benefited over 450 new entrants since 2006. The Academies focus on delivering business skills through a series of short evening courses and have attracted career changers from non-farming backgrounds, as well as the children of farmers. The future may see the development of specialist sector-specific (for example, dairy) academies with strong industry backing (Rickett 2011).

A number of contributors to this book have also made reference to programmes designed to develop and implement succession plans, such as links with the International Farm Transaction Network: Fostering the next generation of farmers in the US as a whole (Goeller, Chapter 9), the work of the Beginning Farmer Centre at Iowa State University (Baker, Chapter 8) and the farm succession education programmes sponsored by the University of Wisconsin Cooperative Extension Service (Kirkpatrick, Chapter 10) and the Farm Succession Aid Programme in Japan (Uchiyama and Whitehead, Chapter 4).

On reflection, therefore, there is much evidence of continued support for family farming, in some cases, providing specific encouragement in terms of the succession process. The efficacy of these approaches, however, is not universally proven where, for example, support is deemed to have been inappropriately targeted towards encouraging and effecting a transfer of the farm and business, which would have taken place without such support. Recognising the complexity of the process, continued intervention, perhaps in more economically challenging times in the short to medium term, will need to be very carefully considered and targeted to achieve maximum effect in any particular circumstance.

Imperatives for Future Generations of Family Farms

'I can't help noticing that farmers are back in fashion!' (Hughes 2010). These are the words of internationally renowned Professor David Hughes, Imperial College, London. In his address, Hughes goes on to identify that, 'in a surprisingly wide range of countries around the globe, I see farmers being featured on supermarket shelves, and in national advertisements for major food brands', a position for farming far removed from that as recent as five years ago (Hughes 2010). Public demands from the farming industry are now seen more positively, perhaps akin to the strength of feeling during and in the two decades post second world war. Food is also most definitely now on the Agenda, with three imperatives clearly articulated in a recent report of the British Cabinet Offic – 'Food Matters' (Cabinet Office 2008) – firstly, the impact of food production techniques on human nutrition, and then, at an international scale, the contribution to the global challenges of climate change and the alleviation of food poverty. Subsequent to this, in the UK the Department of Environment, Food and Rural Affairs (DEFRA) launched the country's first Food Security Assessment (DEFRA 2009a) and has been investigating the merits of locally produced food and the impacts on the environment of encouraging greater consumption of seasonally produced food.

The macro-drivers of the future, at the global level, are endorsed by Giles (2011) as including, on food demand, continuing population growth (predominantly in Asia and Latin America) and expected shifts from low to medium (and high) incomes of increasing proportions of the population in Brazil, India and China. As for food supply, the impacts of changing weather patterns on production (especially in Australia, parts of North America, Asia and Africa), distribution and access to water and other natural resources are also noted as key issues. The future of farming does not solely concern the capacity of resources and practices in global agriculture to meet increasing demands for food. Demands for the provision of a raft of other goods (that is bioenergy and industrial crops) and services (that is conservation and recreational) are also increasing in the context of volatile commodity prices, diminishing non-renewable resources and climate change. Concurrent with such challenges, there is increasing evidence of continued degradation of the soil arising from continued unsustainable intensive agricultural practices in areas of the world including Australia, the US and the UK (Sage 2012).

Indicative of the geographical significance of these concerns, whilst the UK Government (DEFRA) published its 'Vision 2030 – Safeguarding Our Soils: A Strategy for England' (DEFRA 2009b), the 2008 Farm Bill (Food, Conservation, and Energy Act of 2008) was published in the United States. An element of this, the Beginning Farmer/Rancher Development Program endeavours to enhance the food security of the United States by providing beginning farmers and ranchers and their families with the necessary knowledge and skills to make decisions concerning the future sustainable farming of their properties. In Australia, Hamblin (2009) focuses on the policy directions for agricultural land use and the need to farm in more sustainable ways, 'in-keeping with the fragile and unique characteristics of Australian landscapes.' A succession of droughts and water shortages has thrown into clear focus the need for appropriate selection strategies for the provision of the complex array of goods and services in the future (*ibid*). Some family farms will be able to supply a range of public goods and services more easily than others and the typology of family farms outlined above offers some clues as to the likely variable response.

With food security rising on political agendas within the context of increasing demands for land, attention is turning to the comparative advantage of productive agricultural land, in the right location in relation to the market, climate and availability of water. Sage (2012) and others suggest that rising to the challenge of food security will require new ways of thinking in almost every aspect of agricultural production. Lambert et al. (2007) report varied priorities concerning the use of conservation-compatible practices across a range of family farms the US. A proportion of these, the so called 'high sales' businesses, were committed to increasing productivity through practices, such as conservation tillage, that reduce costs and increase yields. There was a clear correlation here between size and importance of farm income, and the human and financial capital invested in the farming practices, although with profit maximisation as the key objective. It is posited, through the typology presented, that there will continue to be a proportion of family farms that will continue to respond to the food security issue by chasing technological developments and opportunities to expand in order to further increase output. These family farms are thus, much more focused on agricultural production, with the traditional attractions to potential successors of mechanisation and farming prowess. The renaissance in agriculture, with demands from all quarters, will undoubtedly influence the minds of potential successors.

The various combinations of physical and human attributes of the farm, as well as a range of exogenous factors, will decide the mix of family farms and their activities in the future. Zasada (2011), for example, identifies that in the peri-urban fringe of Germany the privileged location provides greater scope for multifunctional activity, and many farms have already positioned themselves accordingly. The multifunctional family farm is likely to continue with diversification strengthening its 'offer', with viability supported by the resourcefulness and varied skills of the family and the proximity to the market. Jongeneel et al. (2008), working with family farmers in the Netherlands, suggest however, that multifunctional activity on farms is less likely where 'farm heads' are older, pointing, once again to the continuing importance of effective succession strategies.

In terms of location alone, farm businesses in less favoured areas tend to have less opportunity to diversify the use of available resources. As the *FARMTRANSFERS* rapporteurs have shown, many of these family farms are small, when measured in terms of agricultural output or agricultural area. For these farmers, there is evidence of a positive correlation between the proportion of off-farm work and effective participation in nature conservation activities (Jongeneel et al. 2008). Such small, mainly family based farms are already proving to be important in the conservation of traditional farming practices favouring nature conservation at the landscape scale in the future (Potter and Lobley 1993; Centre for Rural Economics Research, University of Cambridge and CJC Consulting 2002).

Other evidence from the US (Lambert et al. 2007: 87) shows that 'retirement' family farms were more likely to adopt 'conservation compatible practices that save time and effort and do not require major changes in established practices'. They were found to be less likely than larger, so called, 'high sales' farms, to invest in management intensive conservation practices. Retirement and residential farms were also found to be more likely to retire land, a potential for wildlife enhancement.

'Smaller' lifestyle family farms therefore have a continuing role to play in the conservation of farming practice, with benefits to wildlife and landscape, as well as contributing to the character of rural areas and their communities. Hamblin (2009: 1199), elevates the relevance of such benefits, 'who could contemplate a view of Mount Fuji without the rice paddies, or the Alps without the sound of the cow bells?', pointing also to the desire to maintain long established practices for biological and landscape reasons, including for example the park-like *dehesa* system of Spain, the rock gardens of the *maquis* and the Alpine meadows (page 1200).

The challenges to farming and the opportunities provided will vary geographically in nature and degree. In Eastern Europe, they will differ from those of Eastern Australia or the uplands of England. In a recent review of the challenges to rural land management, Hodge (2009: 652) identifies that 'farm businesses need to develop their resilience in the face of greater exposure to the volatilities of world markets and the reduced level of support under agricultural policy', as well as the uncertainties of climate change. Such 'resilience' is the preserve of many family farms and arguments for this are well rehearsed (Pile 1990; Donohue 2001; Price and Evans 2009; Munton 2009). Jones (1996: 197) for instance, refers to the importance of the 'intimate coaxing style of management' of family farms and their strength as 'a long term institution protecting not only its economic base, but also its own place and surrounding'.

What is clear, however, is that despite such resilience and, in many cases, a huge desire for independence, continued targeted support, financial and non-financial, may be needed in order to ease challenges, present and future. The presence of the 'enthusiastic successor', keen to take on the business is paramount to the continuation of the family farming system. One of the purposes of the development of the typology above is to elucidate the different types of family farms and the varied succession routes. As was argued in Chapter 1, this reflects the lack of any single widely applicable definition of family farms but emphasises the diversity of family farming. Equally, policy intervention will vary according to circumstance and should not be based on a 'one size fits all' model. Family farms are heterogeneous in nature and in the means by which to respond to future challenges. As the typology (Figure 13.3) suggests, the need for, the level of, and the appropriate approaches towards support in the future will vary. For the lifestyle/residential family farm, for example, this may entail targeted support to secure specific habitats, landscape and social/community structures recognised as of local, national or international import. In contrast, for the agribusiness family farm assistance is more likely to take the shape of that which facilitates further restructuring, such as improvements to tenure arrangement, to assist with the establishment of the mixture bundle of rights, along with appreciation of the need for growth from the planning system.

In all of this, continuity of management, through close relationships between family members, the 'sharing' of capital assets and the detailed knowledge of the farm resource contribute to the strength of family farms. The successors of the future will have to be highly motivated, skilled in technical and business matters and capable of pre-empting change and planning appropriate responses. Without this, the risk is that the cornerstone of agricultural business across the developed world will fail to meet local, national and global expectations.

References

ADAS Consulting Ltd., University of Plymouth, Queens University Belfast and the Scottish Agricultural College. 2004. *Entry to and Exit from Farming in the United Kingdom*. ADAS Consulting, Wolverhampton. Report to DEFRA, London.

AgraCEAS Consultants. 2003 *Ex Post Evaluation of Measures Under Regulation (EC) No 950/97 on Improving the Efficiency of Agricultural Structures*. Final Report for the European Commission Directorate-General for Agriculture.

Bika, Z. 2007. The Territorial Impact of the Farmers' Early Retirement Scheme. *Sociologia Ruralis*, 47(3): 246.

Bougherara, D. and Latruffe, L. 2010. Potential of the EU 2003 CAP reform on land idling decisions of French landowners: result of a survey of intentions. *Land Use Policy*, 27: 1153–1159.

Cabinet Office. 2008. *Food Matters – Towards a Strategy for the 21st Century*. Cabinet Office, London.

Carruthers, S.P. and Miller, F.A. (eds) 1996. *Crisis on the Family Farm: Ethics or Economics?* Centre for Agricultural Strategy, University of Reading, Reading.

Caskie, P., Davies, J., Campbell, D. and Wallace, M. 2003. *The EU Farmer Early Retirement Scheme: A Tool for Structural Adjustment*. Paper presented at the 77th Annual Conference of the Agricultural Economics Society, University of Plymouth, Seale-Hayne Campus.

Centre for Rural Economics Research, University of Cambridge and CJC Consulting 2002. *Economic Evaluation of Agri-Environment Schemes*. Report to the Department of Environment, Food and Rural Affairs, London.

Department of Environment, Food and Rural Affairs (DEFRA) 2009a. *UK food Security Assessment: Our Approach*. DEFRA, London.

Department of Environment, Food and Rural Affairs (DEFRA) 2009b. *Vision 2030 – Safeguarding Our Soils: A Strategy for England*. DEFRA, London.

Donohue, B. 2001. Reclaiming the Commons. In E.T. Freyfogle (ed.) *The New Agrarianism – Land, Culture and the Community of Life*. Island Press, Washington D.C.

Errington, A.J. 1998. The intergenerational transfer of managerial control in the family farm business: a comparative study of England, France and Canada. *Journal of Agricultural Education and Extension*, 5: 123–136.

Errington, A.J. and Lobley, M. 2002. *Handing Over the Reins: A Comparative Study of Intergenerational Farm Transfers in England, France, Canada and USA*. Paper presented to conference of the Agricultural Economics Society, Aberystwyth, 8–11 April.

Foskey, R. 2002. *Older Farmers and Retirement*. Unpublished report to the Rural Industries Research and Development Corporation, Canberra, ACT.

Gasson, R. 1988. *The Economics of Part-time Farming*. Longman, Harlow.

Gasson, R. and Errington, A. 1993. *The Farm Family Business*. CAB International, Wallingford, p. 304.

Giles, J. 2011. How will the global supply chain look in the future? *Journal of Farm Management*, 14(3): 187–194.

Hamblin, A. 2009. Policy directions for agricultural land use in Australia and other post-industrial economies. *Land Use Policy*, 26: 1195–1204.

Harrison, A. 1996. Family farm support: some basic economic issues. In S.P. Carruthers and F.A. Miller (eds) *Crisis on the Family Farm: Ethics or Economics?* CAS Paper 28. Centre for Agricultural Strategy, Reading.

Hill, B. and Ray, D. 1987. *Economics for Agriculture: Food, Farming and the Rural Economy*. Macmillan, London.

Hodge, I. 2009. Scenarios for rural land management: exploring alternative futures. *Journal of Farm Management*, 13: 633–654.

Hoppe, R.A. and Banker, D.E. 2010. *Structure and Finances of US Farms: Family Farm Report, 2010 Edition*, EIB-66, US Department of Agricultural Economic Research Service, July 2010.

Hughes, D. 2010. *Farmers are Back in Fashion!* [online]. Available at: http://www.profdavidhughes.com/newsblog/latestnews/farmers-are-back-in-fashion/ [accessed 12 July 2011].

Jones, 1996. Family survival and family values. In S.P. Carruthers and F.A. Miller (eds) *Crisis on the Family Farm: Ethics or Economics?* CAS Paper 28. Centre for Agricultural Strategy, Reading.

Jongeneel, R.A., Polman, N.B.P. and Slangen, L.H.G. 2008. Why are Dutch farmers going multifunctional? *Land Use Policy*, 25: 81–94.

Kaine, G.W., Crosby, E.M. and Stayner, R.A. 1997. *Succession and Inheritance on Australian Family Farms*. TRDC Publication no. 198. The Rural Development Centre, University of New England, Armidale.

Kerselaers, E., Rogge, E., Dessein, J., Lauwers, L. and van Huyenbroeck 2011. Prioritising land to be preserved for agriculture: a context-specific value tree. *Land Use Policy*, 28: 219–226.

Kimhi, A. and Bollman, R. 1999. Family farm dynamics in Canada and Israel: the case of farm exits. *Agricultural Economics*, 21: 69–79.

Lambert, D.M., Sullivan, P., Claassen, R. and Foreman, L. 2007. Profiles of US farm households adopting conservation-compatible practices. *Land Use Policy*, 24: 72–88.

Lobley, M., Potter, C., Butler, A., Whitehead, I.R.G. and Millard, N.R. 2005. *The Wider Social Impacts of Changes in the Structure of Agricultural Businesses*. Report to DEFRA, London.

Lobley, M., Baker, J.R. and Whitehead, I.R.G. 2010. Farm succession and retirement: some international comparisons. *Journal of Agriculture, Food Systems, and Community Development*, 1(1): 49–64.

Mazzora, A.P. 2000. Analysis of the evolution of farmers' early retirement policy in Spain. The case of Castille and Leon. *Land Use Policy*, 113–120.

Miljkovic, D. 2000. Optimal timing in the problem of family farm transfer from parent to child: an option value approach. *Journal of Development Economics*, 61: 543–552.

Munton, R. 2009. Rural landownership in the United Kingdom: changing patterns and future possibilities for land use. *Land Use Policy*, 265: 554–561.

Naylor, E. 1982. Retirement policy in French agriculture. *Journal of Agricultural Economics*, 33(1): 25–36.

Newby, H. 1978. *Property, Paternalism and Power: Class and Control in Rural England*. Hutchinson and Co., London.

Newby, H. 1988. *Country Life – A Social History of Rural England*. Sphere Books Ltd., London.

Pile, S. 1990. *The Private Farmer – Transformation and Legitimation in Advanced Capitalist Agriculture*. Dartmouth Publishing Company, Aldershot.

Potter, C. and Lobley, M. 1993. Helping small farms and keeping Europe beautiful: a critical review of the environmental case for supporting the small farm. *Land Use Policy*, 10: 267–279.

Potter, C. and Lobley, M. 1996. Unbroken threads? Succession and its effects on family farms in Britain. *Soiologia Ruralis*, 36, 286–306.

Potter, C. 1998. *Against the Grain: Agri-environmental Reform in the US and EU*. CABI, Wallingford.

Price, L. and Evans, N. 2009. From stress to distress: conceptualising the British family farming patriarchal way of life. *Journal of Rural Studies*, 25: 1–11.

Rickett, A. 2011. Personal communication.

Sage, C. 2012. *Environment and Food*. Routledge, London.

Shoard, M. 1987. *This Land is Our Land – The Struggle for Britain's Countryside*. Paladin, London.

Uchiyama, T., Lobley, M., Errington, A. and Yanagimura, S. 2008. Dimensions of intergenerational farm business transfers in Canada, England, the USA and Japan. *Japanese Journal of Rural Economics*, 10: 33–48.

Ward, N. 1996. Environmental concern and the decline of dynastic family farming. In S.P. Carruthers and F.A. Miller (eds) *Crisis on the Family Farm: Ethics or Economics?* CAS Paper 28. Centre for Agricultural Strategy, Reading.

Whitehead, I.R.G., Errington, A.J.E. and Millard, N.R. (2002). *Economic Evaluation of the Agricultural Tenancies Act 1995*. Report for DEFRA, University of Plymouth, London.

Zasada, I. 2011. Multifunctional peri-urban agriculture – a review of societal demands and the provision of goods and services by farming. *Land Use Policy*, 28: 639–648.

Index

Locators shown in italics refer to tables and figures.